CalcLabs with Mathematica: Single Variable Calculus

FOURTH EDITION

Prepared by

Selwyn Hollis

Armstrong Atlantic State University

BROOKS/COLE
CENGAGE Learning™

Australia • Brazil • Japan • Korea • Mexico • Singapore • Spain • United Kingdom • United States

BROOKS/COLE
CENGAGE Learning

For product information and technology assistance, contact us at
Cengage Learning Customer & Sales Support, 1-800-354-9706

For permission to use material from this text or product, submit all requests online at
www.cengage.com/permissions
Further permissions questions can be emailed to
permissionrequest@cengage.com

ISBN-13: 978-0-495-56063-0
ISBN-10: 0-495-56063-4

Brooks/Cole
10 Davis Drive
Belmont, CA 94002-3098
USA

Cengage Learning is a leading provider of customized learning solutions with office locations around the globe, including Singapore, the United Kingdom, Australia, Mexico, Brazil, and Japan. Locate your local office at:
www.cengage.com/international

Cengage Learning products are represented in Canada by Nelson Education, Ltd.

To learn more about Brooks/Cole, visit
www.cengage.com/brookscole

Purchase any of our products at your local college store or at our preferred online store
www.ichapters.com

Printed in the United States of America
1 2 3 4 5 13 12 11 10 09

ED085

Contents

8 Sequences and Series

9 Projects

Prologue

Over the past two decades the wide availability of powerful computational software such as *Mathematica*® has had a significant effect upon the way calculus is taught and learned. Remarkably, computations that would have been extraordinary just a few years ago—and certainly unimaginable in the times of Newton, Gauss, and Riemann—can be done easily by any student with a laptop computer and *Mathematica*.

The primary goal of this manual is to show you how *Mathematica* can help you learn and use calculus. The approach (we hope) is not to use *Mathematica* as a "black box," but rather as a tool for exploring calculus concepts and the way calculus can be used to solve problems.

Two secondary goals of this manual are: 1) to present in a very concise manner the central ideas of calculus, and 2) to introduce you to many of the capabilities of *Mathematica*. You should be aware of—but hopefully not intimidated by—the fact that *Mathematica* is an enormously complex system that can be frustrating to the beginner. However, with discipline and a little perseverance, you will soon begin to see the basic elegance and (believe it or not) underlying simplicity of *Mathematica*.

The last chapter of this manual contains twenty-four extended exercises, or "projects," which cover a wide range of topics in single-variable calculus. The projects vary considerably in length, level of difficulty, and the amount of guidance provided. We hope that these projects will be interesting—and sometimes fun—while reinforcing important calculus concepts.

Mathematica will not "do calculus" for you. It cannot decide the proper approach to a problem, nor can it interpret results for you. In short, *you* still have to do the thinking, and *you* need to know the fundamental concepts of calculus. You must learn calculus from lectures and your textbook—and most importantly by working problems. That's where *Mathematica* comes in as a learning tool, allowing you to concentrate on concepts and to work interesting problems without getting bogged down in algebraic and computational details.

Until you've had a lot of experience using *Mathematica*, you will need to consult the *Mathematica* Documentation Center frequently. The first thing you should do is familiarize yourself with the Documentation Center, in which you can find detailed descriptions of essentially every element of *Mathematica*, as well as a wealth of tutorial material. Also, the first chapter of this manual is a tutorial that introduces you to essentially all of the basic *Mathematica* concepts necessary for working your way through the remainder of the manual.

■ Web Site

Corrections to this manual, as well as related updates, *Mathematica* notebooks, and add-ons will be posted at www.math.armstrong.edu/faculty/hollis/mmacalc/.

Mathematica 6

This manual was produced entirely with *Mathematica* 6 (6.0.3) and relies upon many of its new features. Hence, any attempt to use this manual with an earlier version of *Mathematica* will only lead to confusion and frustration.

Mathematica 6, which was first released in May of 2007, is a revolutionary new version of *Mathematica*. The sheer number of new features and functions is astonishing. Even *Mathematica* users with years of experience have a vast new world to explore. On the other hand, many of the changes have made using *Mathematica* much simpler and more intuitive. In fact, most commands in this manual have to some degree been shortened, and many significantly so.

The following briefly describes (for those with prior *Mathematica* experience) some of the new ways of doing things that are particularly pertinent to this manual.

Displaying and Combining Graphics

The process of combining graphics has been greatly simplified, since the display of graphics is no longer a "side effect" of the creation of a graphics object. The following describe a few of the most basic consequences.

- Graphics may be suppressed by a semicolon just as any other output. For instance, the following produces no picture or other output.

```
Plot[Sin[x], {x, 0, 2 π}];
```

- Graphics computed inside of `Show` are not displayed individually prior to the output of `Show`.

```
Show[Plot[x^2, {x, -1, 1}],
 Graphics[Circle[{0, .5}, .5]], AspectRatio → Automatic]
```

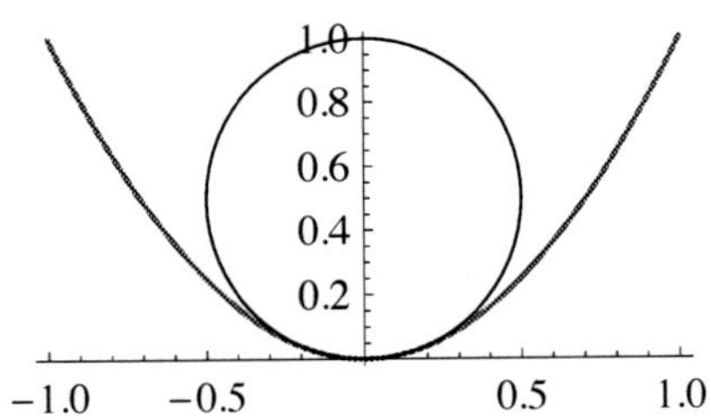

Hence there is no longer any need for the `DisplayTogether` function or the `DisplayFunction` option.

- The function `Graphics` (and `Graphics3D`) produces visual output without the help of `Show`.

```
Graphics[{Gray, Disk[]}, ImageSize → 48]
```

▪ The Standard Packages

Most of the functions that were previously available in the standard packages either have been incorporated into the *Mathematica* kernel or their functionality has been assumed by other functions. The following are a couple of pertinent examples.

- The `Graphics`FilledPlot`` package is gone. `Plot` now has a `Filling` option.

```
Plot[{Sin[x], Sin[2 x], Sin[3 x]},
 {x, 0, 2 π}, Filling → {1 → {3}, 1 → {2}}]
```

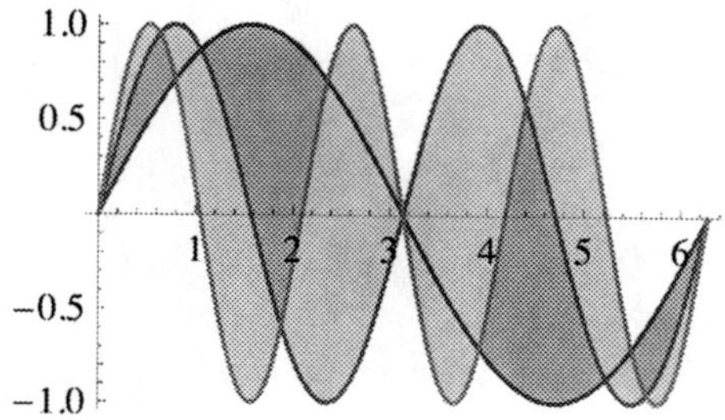

- The `Graphics`ImplicitPlot`` package is gone. `ContourPlot` now works on equations.

```
ContourPlot[x^2 - x y + 2 y^2 == 2, {x, -2, 2}, {y, -2, 2}]
```

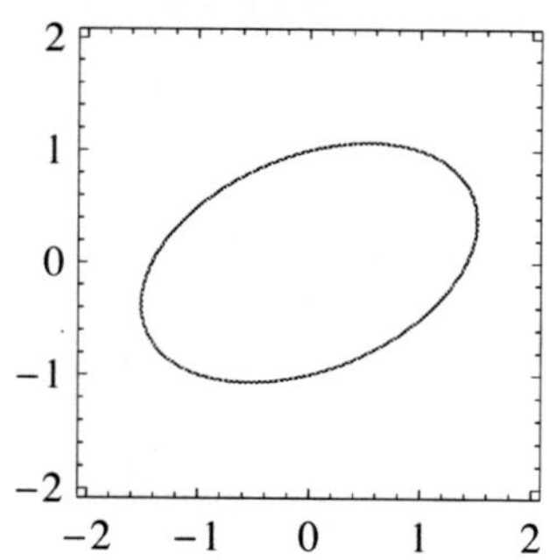

▪ Animate and Manipulate

Making animations is now remarkably simple. `Animate` (which was previously in the `Graphics`Animation`` package) eliminates the need to plot individual frames and collapse the resulting celll group. `Animate` works much like `Table`. (Note that `Table` now *literally* produces a table of graphics.)

```
Table[Plot[x^n, {x, 0, 1}, PlotRange → {0, 1}], {n, 1, 4}]
```

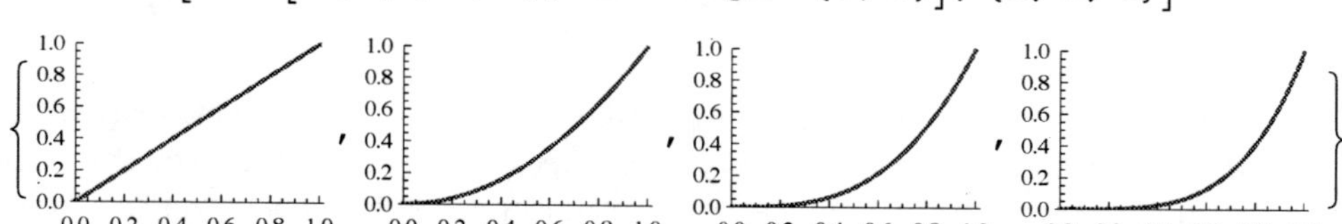

Here's the same command with `Table` replaced by `Animate`.

```
Animate[Plot[x^n, {x, 0, 1}, PlotRange → {0, 1}], {n, 1, 4}]
```

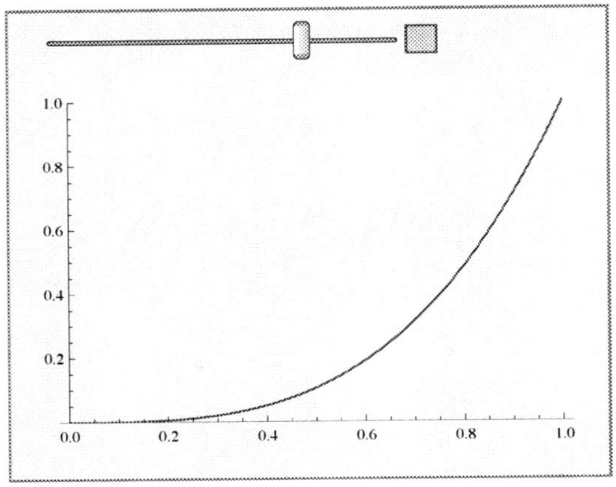

While the default increment for `Table` is 1, `Animate` changes *continuously* by default.

A similar new function is `Manipulate`, which provides a slider (or sliders) for the user to "manually" change one or more parameter values. One can also specify other controls such as radio buttons and pop-up menus, as well as "locators," with which one can manipulate positions of points within the graphic itself. For instance, the following shows the graph of the parabola through three points and allows live manipulation of all three points.

```
Manipulate[
 Plot[InterpolatingPolynomial[{pt1, pt2, pt3}, x],
  {x, -1, 1}, PlotRange → {{-1, 1}, {-1, 1}}],
 {{pt1, {-.5, .5}}, Locator}, {{pt2, {0, -.5}}, Locator},
 {{pt3, {.5, .5}}, Locator}]
```

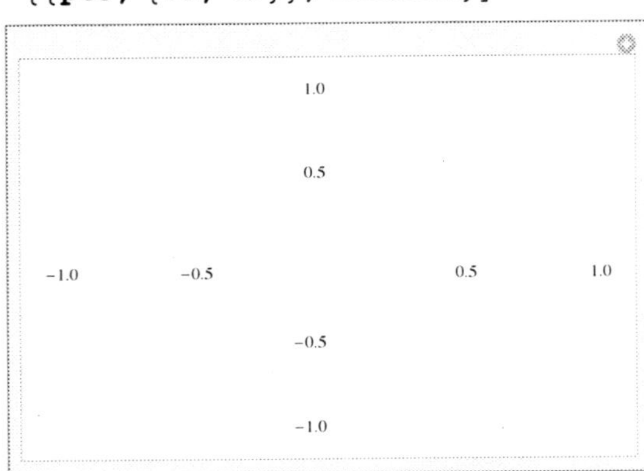

■ The Documentation Center

The old Help Browser has been replaced by the *Documentation Center*, which has greater search capability and the ability to be automatically updated via the internet. The Documentation Center contains a lot of helpful tutorials, but at the time of this publication how to find those tutorials is less than obvious. We hope that a direct link to an index of the tutorials will be added to the "front page" of the Documentation Center in subsequent updates.

1 *Mathematica* Basics

This chapter is an introduction to *Mathematica*. We briefly describe many of the most important and basic elements of *Mathematica* and discuss a few of the more common technical issues related to using *Mathematica*. Since our primary goal is to use *Mathematica* to help us understand calculus, you should not initially spend a great amount of time pouring over the details of this chapter, except as directed by your professor. Simply familiarize yourself with what's here, and refer back to it later as needed.

1.1 Getting Started

Any new user of *Mathematica* must understand several basic facts concerning the user interface, syntax, and the various types of objects that one encounters in using *Mathematica*. This section is a cursory look at some of these fundamentals.

■ The *Mathematica* "Front End"

When you start up *Mathematica*, the first thing you see is a window displaying the contents of a "**notebook**." This window is displayed by *Mathematica*'s **front end**. The front end is the interface between you and the *Mathematica* **kernel**, which does the computations. The following is a typical (simple) notebook in a front end window.

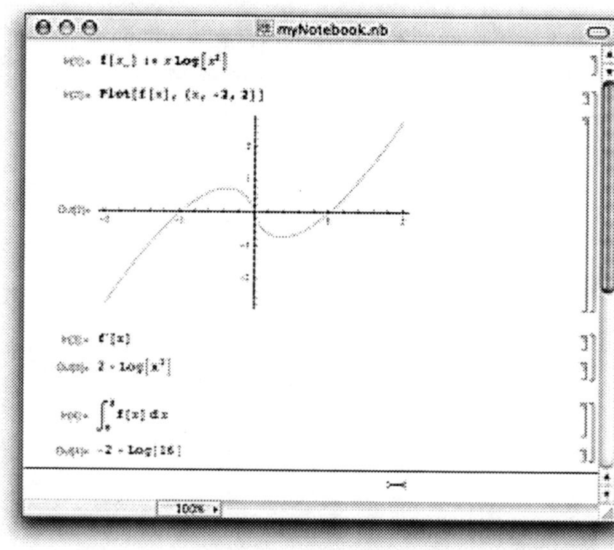

A *Mathematica* notebook is composed of **cells**. On the right side of the window you see **cell brackets**. Each cell in the notebook shown above is either an *input cell*, an *output cell*, or a *graphics cell*. There are several other kinds of cells. Some of these are *text*, *title*, and *section*.

Also notice the horizontal line near the bottom of the window. This indicates the insertion point for the next new cell. To enter a command into a notebook, simply begin typing. The default cell type is *input*. When you're done typing, just press **shift-return** (on a Macintosh, you can also use the "**enter**" key.) To evaluate an existing input cell, simply click anywhere inside the cell (or on the cell bracket) and press **shift-return** (or **enter**).

To create a cell *between* two existing cells, move your cursor over one of the cells toward the other until the "I-beam" cursor becomes horizontal. Then click, and a horizontal line will appear, indicating the desired insertion point. To delete a cell, click on its bracket and then choose Clear from the Edit menu or simply press the DEL key

Palettes. You can enter mathematical expressions so that they appear essentially the same as you would write them on paper or see them in your textbook. For example, to define the

function $f(x) = \sqrt{x^2 + 1}$, we could use the "Input Form"

```
f[x_] := Sqrt[x^2 + 1]
```

or we could use "Standard Form":

$$f[x\_] := \sqrt{x^2 + 1}$$

There is a vast set of keystroke combinations for typing such expressions. However, at first, you will probably want to take advantage of one or more or the standard *palettes* that are available. The image to the right shows the BasicInput palette. Clicking on one of the palette's buttons places the corresponding character/expression at the current input location. This particular palette probably appears by default when you start *Mathematica*, but if not, you can access it or any of the other palettes through your Palette menu.

■ The Documentation Center

Mathematica's Documentation Center may be accessed by selecting Documentation Center from the Help menu. Among the wealth of information available through the Documentation Center are descriptions of all of *Mathematica*'s built-in functions, including examples of their use and links to related tutorials. You can also enter commands from within the Documentation Center. (Whatever you change there will not be saved.)

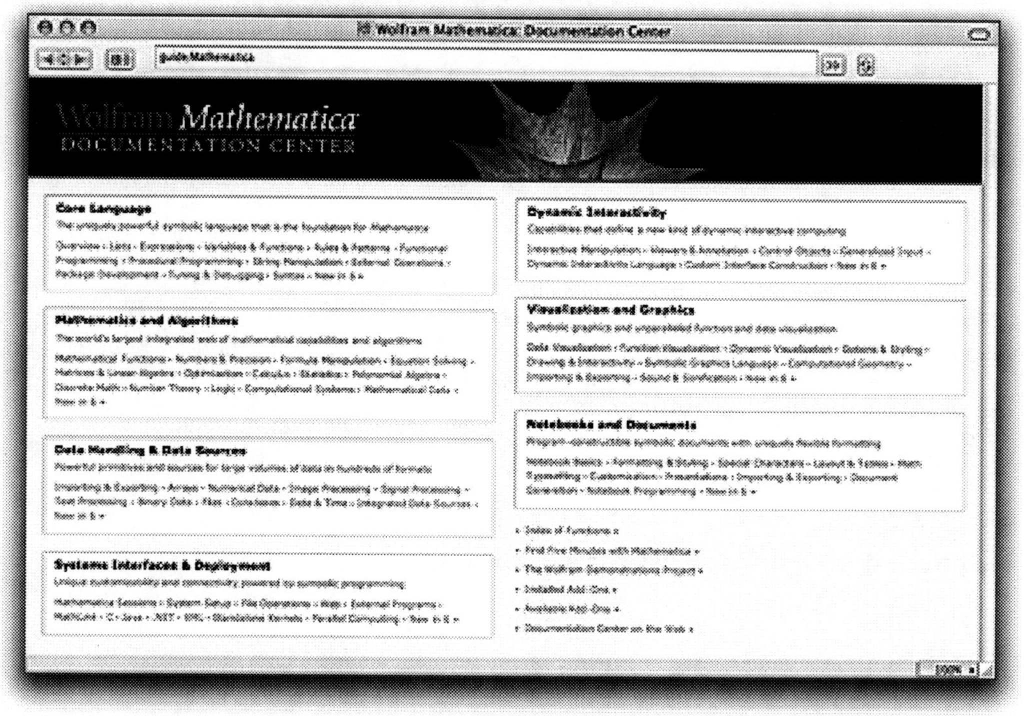

The Documentation Center provides a great deal of tutorial material. If you're a beginner, we suggest that you peruse the tutorials found by entering each of the following in the Documentation Center search field.

tutorial/UsingTheMathematicaSystemOverview

tutorial/InputAndOutputInNotebooksOverview

tutorial/BuildingUpCalculationsOverview

tutorial/EnteringTwoDimensionalExpressionsOverview

tutorial/GraphicsAndSoundOverview

(Hopefully future updates to the Documentation Center will provide a more convenient way to access these and other tutorials.)

■ Basic Calculations

Mathematica allows multiplication to be indicated in three ways. Expressions separated by a space are multiplied. (Note that the system automatically replaces the space with ×.)

```
217 × 5713
```

```
1 239 721
```

An asterisk between expressions indicates multiplication.

```
321 * 5.479
```

```
1758.76
```

When there is no ambiguity, juxtaposed expressions (without spaces) are understood and multiplied.

```
2x
```

```
2 x
```

More than one command can be given in one input cell. A single input cell may consist of two or more lines. A new line within the current cells is obtained by pressing "return." Commands on the same line within a cell must be separated by semicolons. *The output of any command that ends with a semicolon is not displayed.*

```
a = 17 / 13 + 211 / 93;
b = 23 a; c = (a + b) / 51
```

$\frac{34592}{20553}$

The percent sign % refers to the last output (not necessarily the preceding cell).

```
3 / 17 + 1 / 5
```

$\frac{32}{85}$

$\%^2$

$\frac{1024}{7225}$

If the last command on any line is not followed by a semicolon, its result *is* displayed. This effect is very handy for showing intermediate steps in a calculation. The following computes $\frac{25!}{3! \times 22!}(.1)^3\,(.9)^{22}$ (a *binomial probability*).

```
25 !
% / (3 ! × 22 !)
% * .1^3 .9^22
```

```
15 511 210 043 330 985 984 000 000

2300

0.226497
```

💡 You should avoid use of the percent sign as much as possible—especially in separate cells. It is *far* better to give names to results and to use those names in subsequent calculations.

■ Parentheses, Brackets, and Braces

The syntax of *Mathematica* is absolutely strict and consistent (and quite simple once you get used to it). For that reason, there are some differences between *Mathematica*'s syntax and the often inconsistent and sometimes ambiguous mathematical notation that we're all used to. For example:

Parentheses are used *only* for grouping expressions.

```
x (x + 2)^2
```

```
x (2 + x)^2
```

Brackets are used *only* to enclose the argument(s) of a function.

```
Cos[π / 3]
```

$\frac{1}{2}$

Braces are used *only* to enclosed the elements of a *list* (which might represent a set, an ordered pair, or even a matrix).

```
{1, 2, 3, 4}
```

```
{1, 2, 3, 4}
```

Consequently, *Mathematica* does *not* understand what you intend by entering any of these expressions, for example:

```
[x + y (1 - y)]^2
Syntax::tsntxi : "[x + y (1 - y)]" is incomplete; more input is needed.
Syntax::sntxi : Incomplete expression; more input is needed.
```

```
(1, 2)
Syntax::sntxf : "(" cannot be followed by "1, 2)".
Syntax::tsntxi : "1, 2" is incomplete; more input is needed.
```

```
Sin (π)
```

```
π Sin
```

In these first two instances, we were lucky to get an error message. But in the last one, *Mathematica* simply multiplied the expressions π and `Sin`—with no complaint at all!

■ Symbolic vs. Numerical Computation

Computations are typically done symbolically (and therefore *exactly*), unless we request otherwise.

```
123 / √768
```

$\frac{41\sqrt{3}}{16}$

One way to obtain a numerical result is to use the numerical evaluation function, `N`.

```
N[123 / √768]
```

```
4.43838
```

We also get a numerical result if any of the numbers in the expression are made numerical by use of a decimal point.

```
123. / √768
```

```
4.43838
```

Unless we *cause* a numerical result, *Mathematica* typically returns an *exact* form, which in many cases is identical to the expression entered.

$$\text{Cos}\left[\frac{\pi}{12}\right]$$

$$\frac{1+\sqrt{3}}{2\sqrt{2}}$$

```
Log[2]
```

```
Log[2]
```

Names and Capitalization; Basic Functions

All built-in *Mathematica* objects—functions, constants, options, etc.—have full names that begin with a capital letter (or in the case of certain "global" parameters, a dollar sign followed by a capital letter).

```
Sin[π / 3]
```

$$\frac{\sqrt{3}}{2}$$

```
PrimeQ[22 801 763 489]
```

```
True
```

$$\text{Solve}\left[x^2 + x - 12 == 0,\ x\right]$$

```
{{x → -4}, {x → 3}}
```

These full names are used internally by *Mathematica*, even when it is far more natural for us to use a symbolic form such as $x + 7$. `FullForm` lets us see the internal representation of an expression.

```
FullForm[x + 7]
```

```
Plus[7, x]
```

$$\text{FullForm}\left[x == x^2\right]$$

```
Equal[x, Power[x, 2]]
```

When the name of a built-in *Mathematica* object is made of two or more words, all of the component words are capitalized. Some typical *Mathematica*-style names are `FindRoot`, `PlotRange`, `AspectRatio`, `NestList`, etc. In almost all cases the component words are spelled out in full.

All of the familiar "elementary functions" are built-in. In some cases—if you remember to capitalize the first letter and to use brackets instead of braces—you would guess correctly how to use one of those functions. For example,

```
Sin[π / 12]
```

$$\frac{-1+\sqrt{3}}{2\sqrt{2}}$$

There are a few things in this regard that should be pointed out. First, the inverse trigonometric functions use the "arc-function" convention:

```
ArcTan[1]
```

$$\frac{\pi}{4}$$

```
ArcCos[1 / 2]
```

$$\frac{\pi}{3}$$

Also, the natural logarithm is `Log`, not `Ln`.

```
Log[E]
```

1

Algebraic Manipulation

Mathematica is an example of a type of software system that is often called a *computer algebra system*. In addition to numerical computations, a computer algebra system also does *symbolic computation* including the manipulation of algebraic expressions. *Mathematica* has a number of functions for this purpose. Among these are `Expand`, `Factor`, `Together`, and `Apart`.

Expand$\left[(x+5)^3\,(2x-1)^2\right]$

$$125 - 425\,x + 215\,x^2 + 241\,x^3 + 56\,x^4 + 4\,x^5$$

Factor$\left[x^3 + 2\,x^2 - 5\,x - 6\right]$

$$(-2+x)\;(1+x)\;(3+x)$$

Together$\left[x + \frac{2}{x^2+1}\right]$

$$\frac{2+x+x^3}{1+x^2}$$

Apart$\left[\frac{x}{x^2+3\,x+2}\right]$

$$-\frac{1}{1+x}+\frac{2}{2+x}$$

Notice that *Mathematica* does not automatically simplify algebraic expressions:

```
x (3 - x) - 5 x^2 + (x - 1) (2 x + 3)
```

```
(3 - x) x - 5 x^2 + (-1 + x) (3 + 2 x)
```

`Simplify` can be used for this purpose.

```
Simplify[x (3 - x) - 5 x^2 + (x - 1) (2 x + 3)]
```

```
-3 + 4 x - 4 x^2
```

Plotting Graphs: An Introduction to Options

Mathematica is extremely good at creating graphics to help us analyze problems. We will be primarily interested in graphing functions of one variable. This is done with `Plot`.

The function $f(x) = \sin(\pi x (3 - x))$ is graphed on the interval $0 \leq x \leq 3$ as follows.

```
Plot[ Sin[π x (3 - x)], {x, 0, 3}]
```

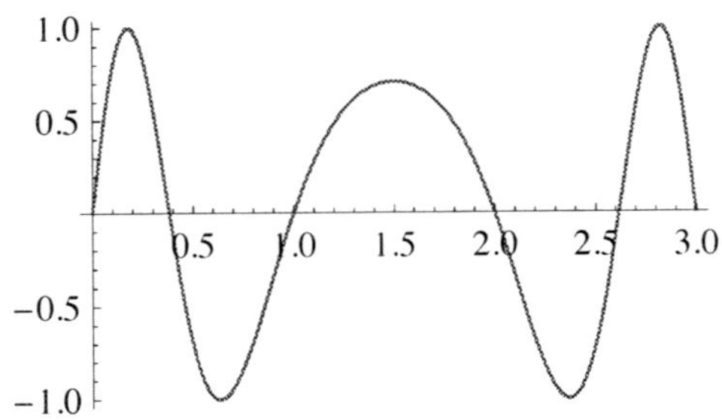

Notice that two *arguments* are provided to `Plot`. The first is our function in the form of an *expression*, and the second is a *list* with three members, specifying (i) the name of the variable, (ii-iii) the left and right endpoints of the interval.

There are numerous ways that we could have affected the appearance of the plot by specifying **options**. Among the options for `Plot` are `PlotRange`, `Ticks`, `AxesLabel`, `AspectRatio`, and `PlotStyle`.

The following creates a plot with labelled axes with no tick marks. Note that the arrow character is typed as [ESC]->[ESC]. (Actually, just -> will do.)

```
Plot[Sin[π x (3 - x)], {x, 0, 3},
 Ticks → {{1, 2, 3}, {-1, 1}}, AxesLabel → {x, y}]
```

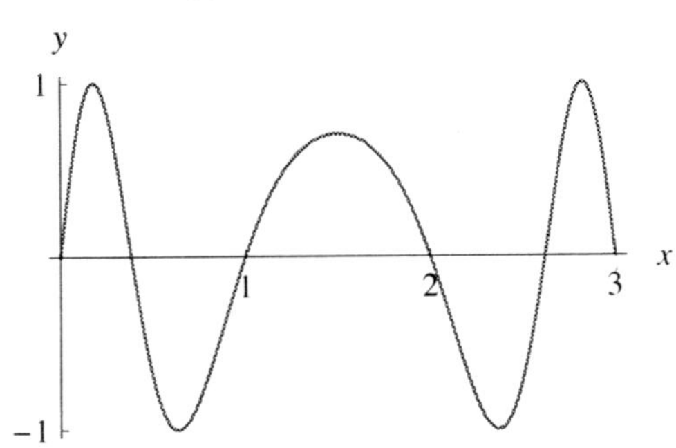

Notice that the following plot chops off high and low parts of the curve.

```
Plot[Sin[2 x^2]/(x^2 + 1), {x, 0, 10}]
```

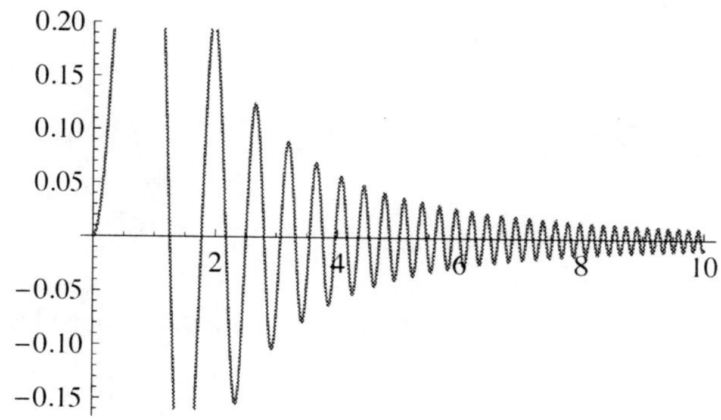

This can be cured with the `PlotRange` option.

```
Plot[Sin[2 x^2]/(x^2 + 1), {x, 0, 10}, PlotRange → All]
```

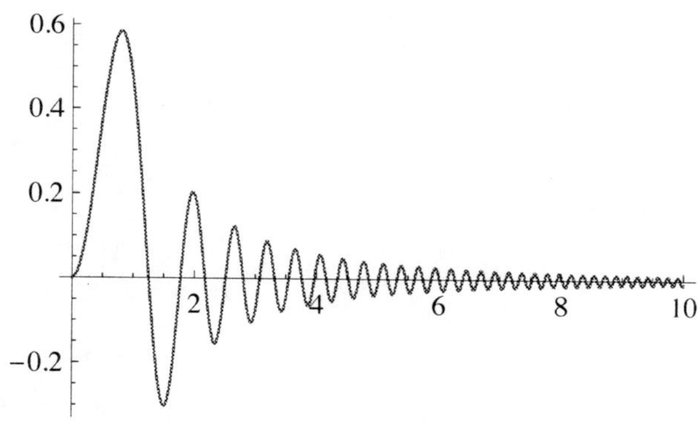

Without our specifying `AspectRatio → Automatic`, the following semicircle would be stretched vertically.

```
Plot[√(1 - x^2), {x, -1, 1}, AspectRatio → Automatic]
```

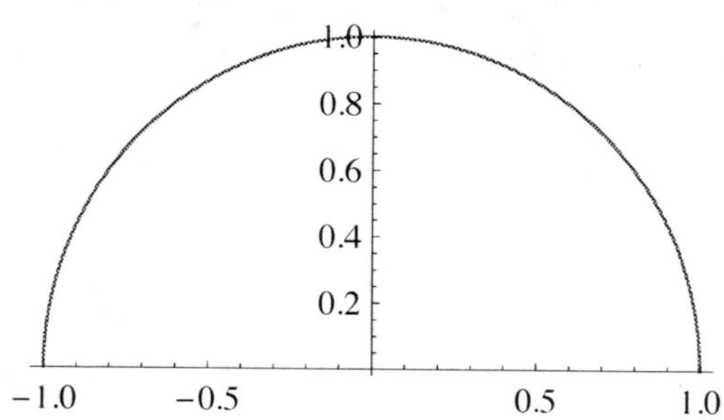

To plot more than one function at once, we give `Plot` a *list* of functions.

```
Plot[{x e^(-x/2) Cos[π x], x e^(-x/2) Sin[π x]}, {x, 0, 12}]
```

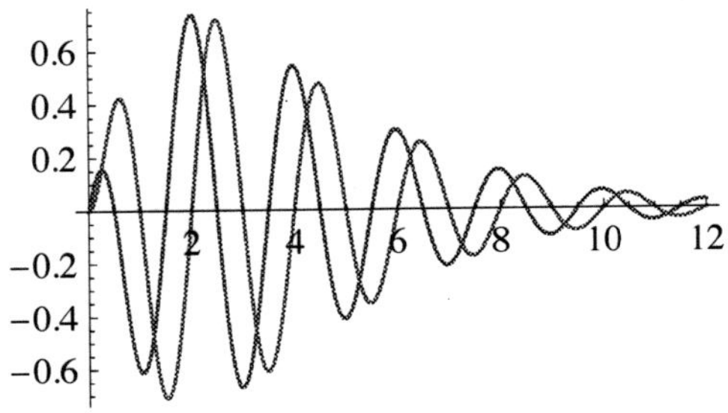

When plotting multiple functions, it is often desirable to plot them with different styles. The `PlotStyle` option lets us do that.

```
Plot[{∛x, x, x^3}, {x, 0, 2}, PlotRange → {0, 1.6},
 PlotStyle → {{Dashing[{.02}]}, {Red}, {Thickness[.007]}}]
```

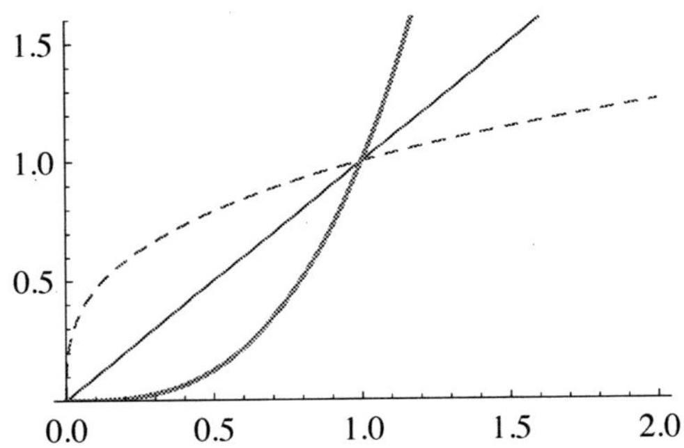

■ Variables

As we mentioned earlier, all built-in *Mathematica* objects begin with an upper-case letter. For that reason, it is usually a good idea to use variable names that begin with a lower-case letter. In this manual we will loosely follow that convention. It is also good practice to give meaningful names to variables and never to make assignments to single letter variables.

Assignments to variables are remembered by *Mathematica* (for the duration of one "kernel session") until the variable is "cleared." *This is probably the single most important thing to remember when you run into difficulties using Mathematica*. (We will have more to say on this in Section 1.8.)

```
piSixths = N[π / 6]

0.523599

Clear[piSixths]; piSixths

0.523599

piSixths
```

Assignments in *Mathematica* are made in two ways: (i) with a single equal sign, or (ii) with a colon followed by an equal sign. For simple assignments such as

```
radius := √(10. / π)
```

it makes little difference which method is used. In this particular case, the consequence of using := is that `radius` has not yet been computed. (Also notice that no output is produced when := is used.) The evaluation has been delayed until we cause it to be done—for example, by entering

```
radius
```

```
1.78412
```

A better example to illustrate delayed evaluation with := is as follows. If we assign a plot to a variable with :=, then no plot is created. The variable is assigned *the command itself*, not the result.

```
graph := Plot[x √(1 - x²), {x, -1, 1}]
```

```
graph
```

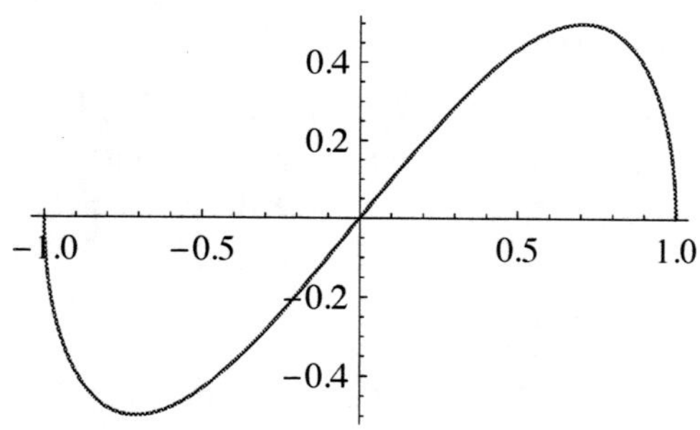

If we assign a plot to a variable with =, then the plot is created and the variable is assigned the resulting *graphics object*.

```
graph = Plot[x √(1 - x²), {x, -1, 1}]
```

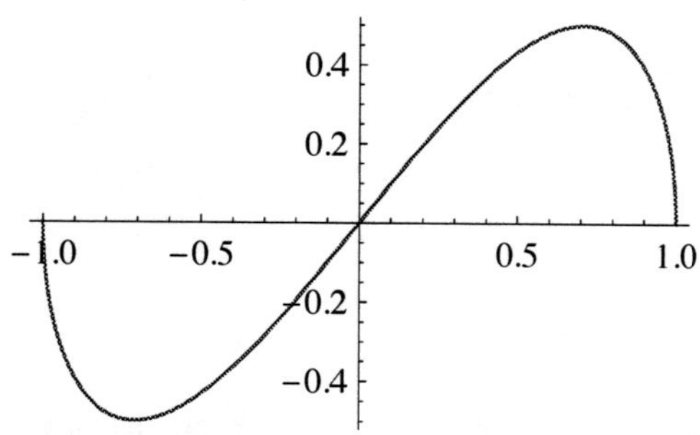

We will address these issues again in Section 1.2. (See also Exercise 9 in this section.)

Tips and Shortcuts

We end this quick tour of *Mathematica* with a few tips and shortcuts with respect to typing expressions.

Entering Exponents, Radicals, and Fractions

To enter an exponent or superscript, press [CTRL]-[6]. (Note that this is analogous to the [SHIFT]-[6] caret (^), which is used for exponentiation in `InputForm`.) To leave the resulting exponent "box," press [→] or [CTRL]-[SPACE].

To enter a subscript, press [CTRL]-[-]. (This is analogous to the [SHIFT]-[-] underscore character (_), which is used to create subscripts in the TEX typesetting language.) To leave the resulting subscript "box," press [→] or [CTRL]-[SPACE].

To enter an expression involving a square root, press [CTRL]-[2]. To leave the resulting square root box, press [→] or [CTRL]-[SPACE].

To enter a fraction, press [CTRL]-[/]. To move from the numerator box to the denominator box, press [TAB] To exit the fraction, press [→] or [CTRL]-[SPACE].

Greek Letters and Other Special Characters

Many special characters and symbols can be typed easily by pressing the [ESC] key before and after typing some easily remembered standard character(s). For instance, to type the Greek letter α (alpha), just type [ESC]`a`[ESC] or [ESC]`alpha`[ESC]. Other Greek letters can be typed similarly.

Other shortcuts for common special characters include:

[ESC]`ee`[ESC] produces the constant e, the base of the natural logarithm.

[ESC]`ii`[ESC] produces the imaginary number $i = \sqrt{-1}$.

[ESC]`int`[ESC] produces an integral sign $\int$.

[ESC]`pd`[ESC] produces a derivative operator ∂.

[ESC]`inf`[ESC] produces an infinity symbol ∞.

Of course, you may prefer to use the buttons on the BasicInput palette.

We should also point out that the familiar numbers denoted by e, i, and π can be entered as such or as `E`, `I`, and `Pi`, respectively.

```
{ⅇ == E, ⅈ == I, π == Pi}
```

```
{True, True, True}
```

Be careful to notice, however, that an ordinary e is not the same as e, and an ordinary i is not the same as i.

◆ Exercises

1. Compute both an exact and a numerical value for each of the following numbers.

   a) $\left(23^3 - 3\,(117 - 48)^2\right) \Big/ \sqrt{7^5 - 5^7}$ b) $\cos \frac{319\,\pi}{12}$ c) $\frac{83!}{111!}$ d) $\ln 2981$

2. Use `Simplify` on each of the following.

   a) $\ln\left(2e^5\right)$ b) $1 + \cos 2x$ c) $\dfrac{x + (x\,(x - 1))^3 - 4}{x^2 + x - 6}$

3. Factor each of these polynomials:

   a) $6x^3 + 47x^2 + 71x - 70$ b) $12x^6 - 56x^5 + 100x^4 - 80x^3 + 20x^2 + 8x - 4$

4. Plot the function $f(x) = \dfrac{\sin(x^3)}{x^3}$ on the interval $0 \le x \le 1$ with:

   a) no options   b) `PlotRange→All`

   c) `PlotRange→All, AspectRatio→Automatic`

5. Create a plot containing the graphs of $y = x^2$ and $y = x^5$ over $0 \le x \le 2$ with:

   a) no options   b) `PlotRange→All`   c) `PlotRange→{0,2}`

   d) `PlotStyle→{Red,Blue}`   e) `PlotStyle→{{Red,Thick},{Blue,Thick}}`

6. Look up each of the following functions in the Documentation Center, and then plot them on the indicated interval.

   a) `Floor`, $0 \le x \le 10$ b) `PrimePi`, $0 \le x \le 100$

7. Use the Documentation Center to learn what `RandomReal[]` and `RandomInteger[]` do. Then enter and the following and explain the output:

```
GraphicsRow[{Plot[RandomReal[] x, {x, 0, 1}],
  Plot[RandomInteger[] x, {x, 0, 1}]}]
```

8. `RandomReal` and `RandomInteger` provide a good illustrations of the difference between using = and using `:=` in an assignment. Enter each of the following several times. Describe and explain the difference in the results.

```
r = RandomInteger[100]; Table[r, {10}]
```

   and

```
r := RandomInteger[100]; Table[r, {10}]
```

1.2 Functions

■ Defining Functions

■ Blank

When defining a function, it is essential to follow each argument by a `Blank` (or "underscore"). Also, recall that the arguments of a function are enclosed by brackets. For example, we would define the function $f(x) = x^3 - 2$ by entering

```
f[x_] := x^3 - 2 x
```

Then we can evaluate the function at any number

```
f[3]
```

```
21
```

or plot its graph:

```
Plot[f[x], {x, -2, 2}]
```

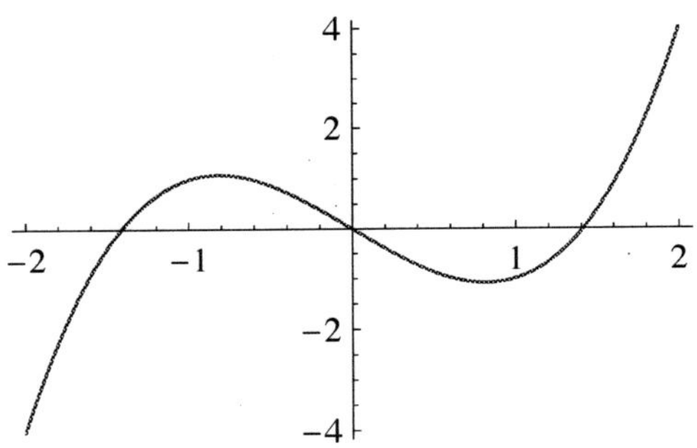

⚠ Note that the `Blank` appears next to `x` *only on the left side of the expression*. Also, when defining a function with more than one argument, a `Blank` must follow each one. For instance:

```
g[x_, y_, z_] := x + y + z
g[1, 2, 3]
```

```
6
```

■ Set (=) versus SetDelayed (:=)

Definitions of functions—and assignments of expressions to variables in general—can be made using either "equal" or "colon-equal." The difference between these two ways is described by the full *Mathematica* names of the = and `:=` symbols, which are `Set` and `SetDelayed`.

When an assignment is made using `Set`, any calculations that are indicated on the right side are done as the assignment is entered. When an assignment is made using `SetDelayed`, any calculations that are indicated on the right side are delayed until the defined expression is used.

In many cases, such as in the definitions of `f` and `g` above, it makes no difference which is used. To see a simple example that indicates the importance of using `Set` rather than `SetDelayed`, let's suppose we want to define $f(x)$ to be the derivative of $(x + 1)\cos x$. If we enter

```
f[x_] := ∂x ((x + 1) Cos[x])
```

notice what happens when we try to evaluate $f(2)$:

```
f[2]
General::ivar: 2 is not a valid variable. ≫
∂2 (3 Cos[2])
```

However, if we enter

```
f[x_] = ∂x ((x + 1) Cos[x])
Cos[x] - (1 + x) Sin[x]
```

then f works the way we want it to:

```
f[2]
Cos[2] - 3 Sin[2]
```

When is it important to use `SetDelayed` rather than `Set`? Here's a simple example: Suppose that we want to plot the graph of $\sin kx$ for several values of k and that we decide to define a function as follows to create each of the desired plots.

```
plotSin[k_] := Plot[Sin[k x], {x, 0, 2 π}]
```

For instance, with $k = 2$, we get the following graph:

```
plotSin[2]
```

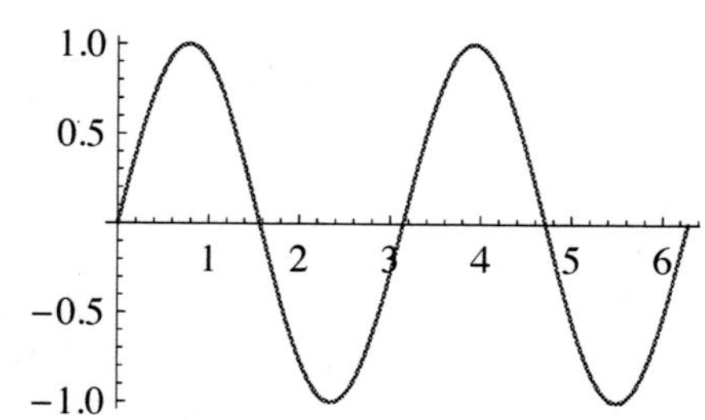

But what would have happened if we used = rather than `:=` in the definition of `plotSin`? *Mathematica* would attempted to plot the graph of $\sin kx$ immediately when the definition is entered. This doesn't work because k has no numerical value, and an empty plot results.

```
plotSin[k_] = Plot[Sin[k x], {x, 0, 2 π}, PlotRange → All]
```

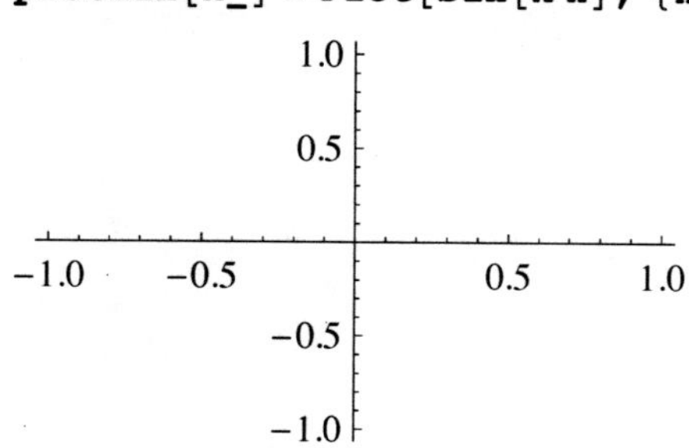

The empty plot that was produced and assigned to `plotSin[k_]` will now be the "value" of `plotsin[k]` for any k; for instance:

```
plotSin[2]
```

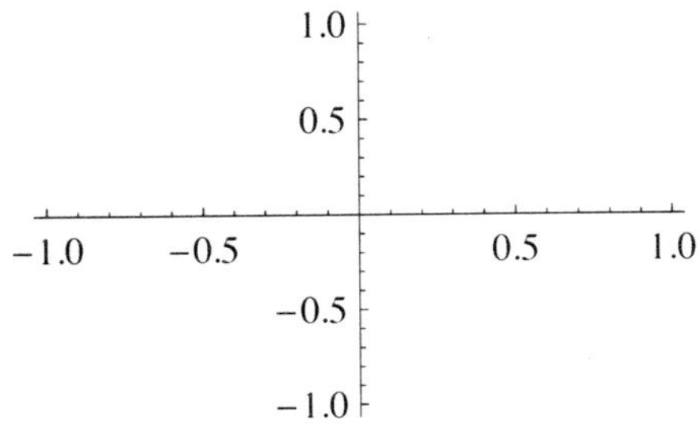

■ Applying Functions with @ and //

Suppose that we define a simple function such as

```
f[x_] := x (x - 1)
```

Naturally, we could evaluate this function at, say, $x = 3$, by entering

```
f[3]
```

```
6
```

There are two other ways to do the same thing. We can evaluate f *at* 3 by entering

```
f@3
```

```
-1
```

On the other hand, we can apply f to 3 like this:

```
3 // f
```

```
6
```

We will use this *postfix* method of function application frequently, often for the purpose of applying either `Simplify` to some expression or `N` to some numerical calculation. For example, we will use the following style when doing a symbolic calculation:

```
2 x + x (5 x + 1) // Simplify
```

```
x (3 + 5 x)
```

When doing an exact numerical calculation, we will commonly use a style that is similar but displays the exact value followed by the numerical value:

$\sqrt{585} \Big/ 33$

```
% // N
```

$\frac{\sqrt{65}}{11}$

```
0.732933
```

■ Piecewise-defined Functions

Mathematica has two primary logical functions that we can use to enter definitions of piecewise- defined functions. These are `If` and `Which`. `If` usually works best for functions with two pieces, such as

$$f(x) = \begin{cases} x, & \text{if } x \leq 1; \\ x-2, & \text{if } x > 1. \end{cases}$$

This function can be entered and plotted as follows.

```
f[x_] := If[x ≤ 1, x, x - 2]
Plot[f[x], {x, -1, 3}]
```

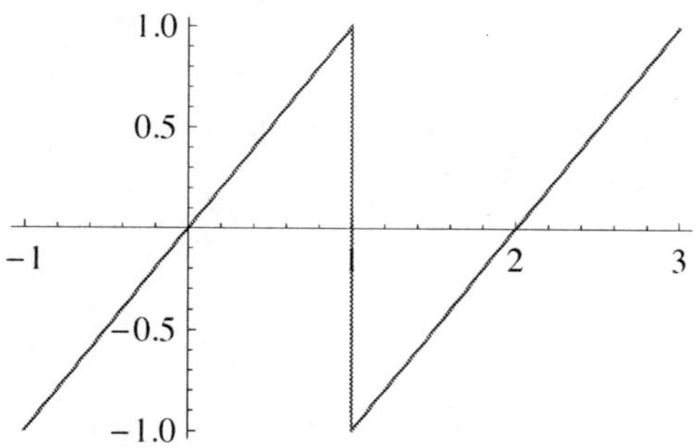

More complicated functions are better handled with `Which`. For example, the function

$$f(x) = \begin{cases} 1, & \text{if } x \leq -1 \\ -x, & \text{if } -1 < x \leq 1 \\ -1, & \text{if } x > 1 \end{cases}$$

can be entered and plotted as follows.

```
f[x_] := Which[x ≤ -1, 1, -1 < x ≤ 1, -x, x > 1, -1]
Plot[f[x], {x, -3, 3}]
```

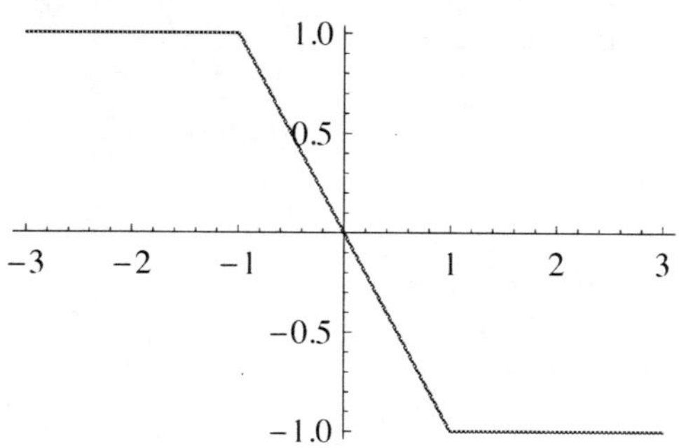

■ Piecewise

`Piecewise` was new in *Mathematica* 5.1 and is similar to `Which`. While `Which` must be supplied with an alternating sequence of conditions and values, *i.e.*,

`Which[` $cond_1, val_1, cond_2, val_2, \ldots$ `]`,

`Piecewise` takes a single list containing value-condition *pairs*, as in

`Piecewise[` $\{\{val_1, cond_1\}, \{val_2, cond_2\}, \ldots\}$ `]`

For example, the function

$$f(x) = \begin{cases} 1, & \text{if } x \le -1 \\ -x, & \text{if } -1 < x \le 1 \\ -1, & \text{if } x > 1 \end{cases}$$

can be input as follows. Notice the automatic display of the output in "left-bracket form."

```
f[x_] = Piecewise[{{1, x ≤ -1}, {-x, -1 < x ≤ 1}, {-1, 1 < x}}]
```

$$\begin{cases} 1 & x \le -1 \\ -x & -1 < x \le 1 \\ -1 & 1 < x \end{cases}$$

It is also possible to enter such a definition in the same left-bracket form as follows: Press ESC pw ESC followed by CTRL RET. Then you'll have a template like this:

$$\begin{cases} \square & \square \\ \square & \square \end{cases}$$

Pressing CTRL RET again will add another row.

■ Abs, Floor, and Mod

There are a handful of built-in piecewise-defined functions, including `Abs`, `Floor`, and `Mod`, whose graphs are shown below.

```
GraphicsRow[
 {Plot[Abs[x], {x, -1, 1}], Plot[Floor[x], {x, -2, 3}],
 Plot[Mod[x, 1], {x, 0, 3}]}]
```

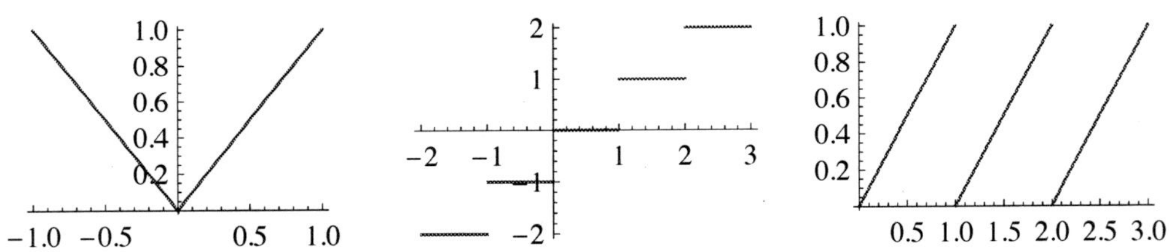

The following functions are built with `Abs`.

```
GraphicsRow[{Plot[Abs[Cos[x] - .5], {x, 0, 4 π}],
  Plot[Abs[Abs[x] - 1], {x, -3, 3}]}]
```

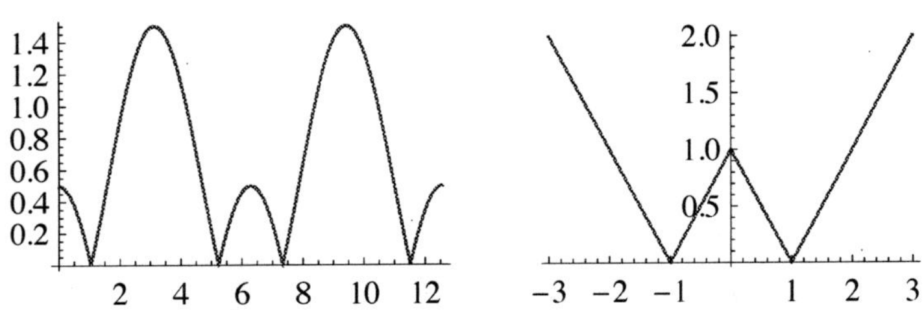

`Floor` helped create each of these *step functions*:

```
GraphicsRow[{Plot[(-1)^Floor[x], {x, -2, 3}],
  Plot[Sin[π Floor[6 x] / 6], {x, 0, 2}]}]
```

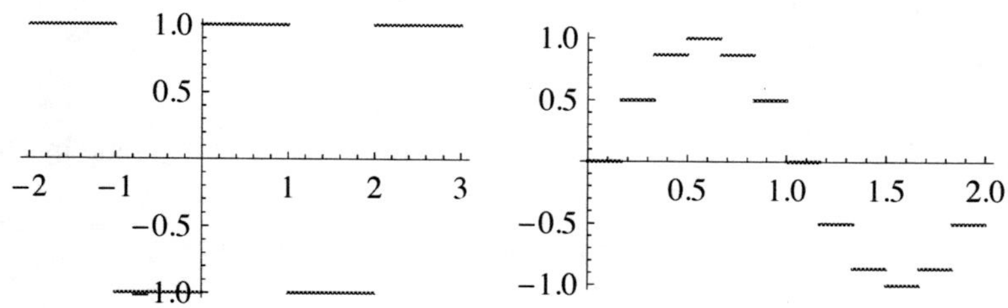

A *periodic function* based on a piece of any graph can be constructed with `Mod`.

```
GraphicsRow[{Plot[Sin[π Mod[x, 1] / 2], {x, 0, 3}],
  Plot[Abs[Mod[x, 2] - 1], {x, -4, 4}]}]
```

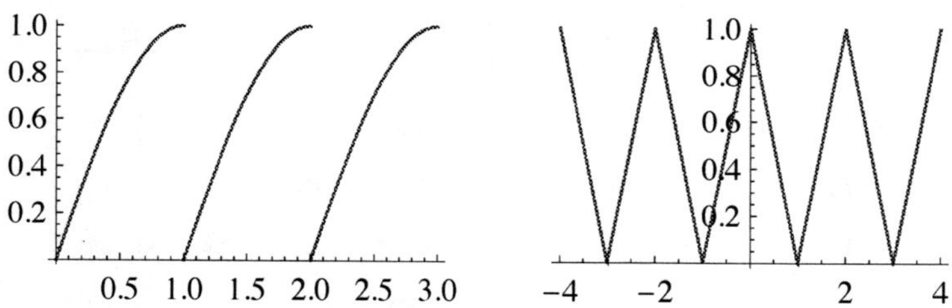

◆ Exercises

1. Enter definitions for each of $f(x) = x\,(x-2)^2$, $g(x) = x - 3$, and $h(x) = x^2 - 1$ and then compute and simplify each of the following:

   a) $f(g(h(x)))$ b) $h(g(f(x)))$ c) $f(h(g(x)))$

2. Enter a definition for $f(x) = \left(1 + x^2\right)^{-1}$. Then create a plot (over $-3 \le x \le 3$) showing the graphs of:

   a) $f(x), f(x-1)$, and $f(x+1)$ b) $f(x), f(2x)$, and $f(8x)$

3. Using `If`, define and plot each of a) $f(x) = \begin{cases} -x, & \text{if } x < 0 \\ x^2 - 1, & \text{if } x \ge 0 \end{cases}$ b) $f(x) = \begin{cases} 1, & \text{if } x \le \pi/2 \\ \sin x, & \text{if } x > \pi/2 \end{cases}$

4. Using `Which`, define and plot each of

   a) $f(x) = \begin{cases} 1, & \text{if } x < 0 \\ 1 - x^2, & \text{if } 0 \le x \le 2 \\ -3, & \text{if } x > 2 \end{cases}$ b) $f(x) = \begin{cases} x + 1, & \text{if } x < 0 \\ 1 - 2x, & \text{if } 0 \le x \le 1 \\ x - 2, & \text{if } x > 1 \end{cases}$

5. Redo Exercise 4 with `Piecewise`.

6. Plot each of the functions:

   a) $f(x) = (x - \text{Floor}(x))^2$ b) $f(x) = x\,(-1)^{\text{Floor}(x)}$

7. The function `Mod` provides an easy way to create a *periodic* function from a simpler function that describes a single period. For example, the function

```
f0[x_] := √(1 - (x - 1)²)
```

   describes the top half of the circle with radius 1 centered at (1, 0). Plot its graph by entering

```
Plot[f0[x], {x, 0, 2},
 AspectRatio → Automatic, PlotRange → {-.5, 1.5}]
```

   The corresponding periodic function with period 2 is

```
f[x_] := f0[Mod[x, 2]]
```

   In a similar manner, define and plot (for $0 \le x \le 9$) a periodic function with period 3 that agrees with $f_0(x) = e^{-x}$ on the interval $0 \le x < 3$.

8. The *unit step function* is defined by $u(x) = \begin{cases} 0, & x < 0, \\ 1, & x \ge 0. \end{cases}$ Its *Mathematica* name is `UnitStep`.

   a) Using `UnitStep`, plot the graph of u on $-1 \le x \le 2$.

   b) Plot $f(x) = u(x-1)$ on $-1 \le x \le 2$ and express $f(x)$ in piecewise form.

   c) Plot $f(x) = u(x-1)\sin(2\pi x)$ on $-1 \le x \le 2$ and express $f(x)$ in piecewise form.

   d) Plot $f(x) = u(x-2) - h(x-1)$ on $0 \le x \le 3$ and express $f(x)$ in piecewise form.

   e) Plot $f(x) = (u(x-2) - u(x-1))\sin(2\pi x)$ on $0 \le x \le 3$ and express $f(x)$ in piecewise form.

1.3 Equations

Certainly one of the most frequent mathematical tasks that we need to do is to solve an equation. In order to be able to solve equations with *Mathematica*, we first need to understand how equations are formed. The important thing to remember is that *double equal signs* are used to form an equation.

Actually, double equal signs constitute a *logical test* that returns either `True`, `False`, or the equation itself.

```
2 * 17 - 34 == 0
```

```
True
```

```
3 == 4
```

```
False
```

```
x² - 4 == 0
```

```
-4 + x² == 0
```

Notice, however, that *Mathematica* returns `True` only when the expressions on each side are identical. Only the most superficial simplification is done prior to the test, as in:

2 x + x - 2 == 3 x + 5 - 7

`True`

Notice that for this equation:

$$\mathbf{x^2 - 4 == (x + 2)\ (x - 2)}$$

$$-4 + x^2 == (-2 + x)\ (2 + x)$$

a nontrivial operation must be done to one side or the other before the expressions become truly identical. Finally, notice that *Mathematica* returns an error message if we use a single equal sign improperly.

3 = 4

Set::setraw: Cannot assign to raw object 3. ≫

4

The single equal sign is used only for *assignments* such as

$$\mathbf{area = \pi\, r^2}$$

$$\pi\, r^2$$

Mathematica has three functions for solving ordinary equations. These are `Solve`, `NSolve`, and `FindRoot`. `Solve` works very well on polynomial and many other algebraic equations.

$$\mathbf{Solve\left[6\,x^3 - 23\,x^2 + 25\,x - 6 == 0,\ x\right]}$$

$$\left\{\left\{x \to \frac{1}{3}\right\},\ \left\{x \to \frac{3}{2}\right\},\ \{x \to 2\}\right\}$$

$$\mathbf{Solve\left[x^4 - 2\,x^3 - x^2 + 6\,x - 6 == 0,\ x\right]}$$

$$\left\{\{x \to 1 - i\},\ \{x \to 1 + i\},\ \left\{x \to -\sqrt{3}\right\},\ \left\{x \to \sqrt{3}\right\}\right\}$$

$$\mathbf{Solve\left[\sqrt{x - 1} + x == 4 + \sqrt{x + 4},\ x\right]}$$

$$\{\{x \to 5\}\}$$

`Solve` will also give solutions to trigonometric (or exponential/logarithmic) equations, but frequently gives a warning.

$$\mathbf{Solve\left[2\,Sin[2\,x] == Cos[x]^2 - 1,\ x\right]}$$

Solve::ifun :

Inverse functions are being used by Solve, so some solutions may not be found; use Reduce for complete solution information. ≫

$$\left\{\{x \to 0\},\ \{x \to -\pi\},\ \{x \to \pi\},\ \left\{x \to \mathrm{ArcCos}\left[-\frac{1}{\sqrt{17}}\right]\right\},\ \left\{x \to -\mathrm{ArcCos}\left[\frac{1}{\sqrt{17}}\right]\right\}\right\}$$

`Solve` will also find solutions of a system of equations. The equations must be given as elements of a list, *i.e.*, separated by commas and enclosed in braces.

```
Solve[{x^2 - y == 1, -x + y == 1}, {x, y}]
```

```
{{y → 0, x → -1}, {y → 3, x → 2}}
```

The solutions of polynomial equations of degree five or greater generally cannot be found in any exact form. Notice how *Mathematica* "avoids" the problem:

```
Solve[x^5 - 10 x^2 + 5 x + 1 == 0, x]
```

```
{{x → Root[1 + 5 #1 - 10 #1^2 + #1^5 &, 1]},
 {x → Root[1 + 5 #1 - 10 #1^2 + #1^5 &, 2]},
 {x → Root[1 + 5 #1 - 10 #1^2 + #1^5 &, 3]},
 {x → Root[1 + 5 #1 - 10 #1^2 + #1^5 &, 4]},
 {x → Root[1 + 5 #1 - 10 #1^2 + #1^5 &, 5]}}
```

In such situations, we can always resort to numerical solutions. `NSolve` finds a numerical approximation to each solution of a polynomial equation, including complex solutions.

```
NSolve[x^5 - 10 x^2 + 5 x + 1 == 0, x]
```

```
{{x → -1.22065 - 1.89169 i}, {x → -1.22065 + 1.89169 i},
 {x → -0.153102}, {x → 0.66946}, {x → 1.92494}}
```

In many situations where `Solve` *is* successful, such as:

```
Solve[x^3 - 10 x^2 + 5 x + 1 == 0, x]
```

$$\left\{\left\{x \to \frac{10}{3} + \frac{85}{3\left(\frac{1}{2}\left(1523+9\,i\sqrt{1691}\right)\right)^{1/3}} + \frac{1}{3}\left(\frac{1}{2}\left(1523+9\,i\sqrt{1691}\right)\right)^{1/3}\right\},\right.$$

$$\left\{x \to \frac{10}{3} - \frac{1}{6}\left(1+i\sqrt{3}\right)\left(\frac{1}{2}\left(1523+9\,i\sqrt{1691}\right)\right)^{1/3} - \frac{85\left(1-i\sqrt{3}\right)}{3\,2^{2/3}\left(1523+9\,i\sqrt{1691}\right)^{1/3}}\right\},$$

$$\left.\left\{x \to \frac{10}{3} - \frac{1}{6}\left(1-i\sqrt{3}\right)\left(\frac{1}{2}\left(1523+9\,i\sqrt{1691}\right)\right)^{1/3} - \frac{85\left(1+i\sqrt{3}\right)}{3\,2^{2/3}\left(1523+9\,i\sqrt{1691}\right)^{1/3}}\right\}\right\}$$

it may still be preferable to use `NSolve`:

```
NSolve[x^3 - 10 x^2 + 5 x + 1 == 0, x]
```

```
{{x → -0.152671}, {x → 0.692369}, {x → 9.4603}}
```

Many equations *require* the use of `FindRoot`, which incorporates a numerical procedure. For example, consider

$$x^2 = \cos x.$$

`FindRoot` can find only one solution at a time and requires us to supply an initial guess at the solution we're looking for. An appropriate initial guess can usually be determined by examining a graph.

```
Plot[{x^2, Cos[x]}, {x, 0, 1}]
```

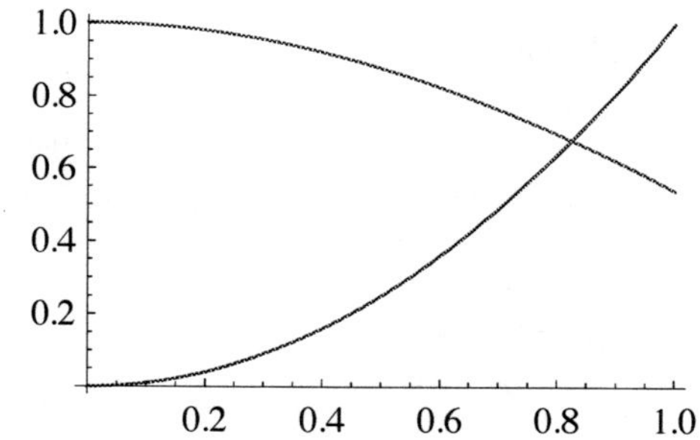


```
FindRoot[x^2 == Cos[x], {x, .85}]

{x → 0.824132}
```

◆ Exercises

1. The equation $x^3 = x + 1$ has one real solution. Find its exact value with `Solve` and its numerical value with `NSolve`.

2. Use `Solve` to find the solution(s) of each of the systems:

   a) $2x + 3y = 1,\ x^2 + y^2 = 1$ b) $3x - 2y = 5,\ 7x + 3y = 2$

   c) $x + y + z = 2,\ x - y + z = 1,\ x^2 + y^2 + z^2 = 2$

3. Use `Solve` on the *underdetermined* system

   $$x + y + 2z = 2,\quad x - 2y + z = 1$$

   and interpret the result. Try each combination of *solve variables*: `{x,y,z}`, `{x,y}`, `{x,z}`, and `{y,z}`. Which gives the "cleanest" solution?

4. For each of the following functions, plot the graph to determine the approximate location of each of its zeros. Then find each of the zeros with `FindRoot`.

   a) $f(x) = e^{-x/2}$ b) $f(x) = x - 9\cos x$ c) $f(x) = x^2 - \tan^{-1} x$

5. For each of the following equations, plot both sides of the equation to determine the approximate location of each of its solutions in the specified interval. Then find each of the solutions with `FindRoot`.

   a) $\sin x \cos 2x = \cos x \sin 3x,\ 0 \le x \le 2\pi$ b) $\sin x^2 = \sin^2 x,\ 0 \le x \le \pi$

   c) $\tan x = x,\ 0 \le x \le 3\pi$ d) $\sin x = e^{-x},\ 0 \le x \le 4$

6. Two spheres have a combined volume of 148 cubic inches and a combined surface area of 160 square inches. Find the radii of the two spheres.

7. Two spheres have a combined volume of 148 cubic inches and a combined surface area of 160 square inches. Find the radii of the two spheres.

8. An open-topped aquarium holds 40 cubic feet of water and is made of 60 square feet of glass. The length of the aquarium's base is twice its width. Find the dimensions of the aquarium.

9. a) Find the equation of the parabola that passes through the points $(-1, 1)$, $(1, 2)$, and $(2, 3)$.

   b) Find the cubic polynomial $f(x)$ such that $f(1) = f(2) = f(3) = 1$ and $f(4) = 7$.

1.4 Lists

Lists are ubiquitous in *Mathematica*. A list is anything that takes the form of a series of objects separated by commas and enclosed in braces, such as:

```
{a, b, c}
{{1, 3}, {2, 5}}
{x, 1, 2}
```

$\{x^2 + y == 2,\ 2\,x - y == 0\}$

Many built-in commands expect lists for certain arguments. For example, in

```
Plot[Cos[π Sin[x]], {x, 0, 2 π}]
```

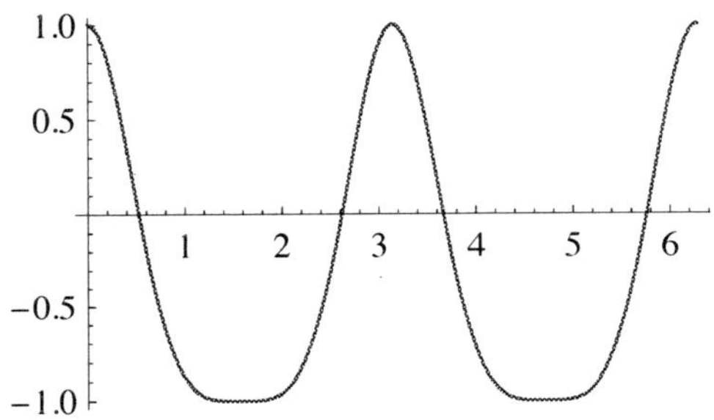

the second argument is a list that specifies the variable name and the interval over which to plot. In

$\texttt{Solve}[\{x^2 + y == 2,\ 2\,x - y == 0\},\ \{x, y\}]$

$\{\{y \to 2\,(-1 - \sqrt{3}),\ x \to -1 - \sqrt{3}\},\ \{y \to 2\,(-1 + \sqrt{3}),\ x \to -1 + \sqrt{3}\}\}$

each of the two arguments is a list, and the result is also a list (of lists).

■ Listable Functions

Most built-in *Mathematica* functions and operations are *listable*. When a listable function is applied to a list, it is applied to each element of the list and returns the result as a list. For example,

$\{1, 2, 3, 4, 5, 6, 7, 8, 9\}^2$

```
{1, 4, 9, 16, 25, 36, 49, 64, 81}
```

$\sqrt{\{\{\{2, 3\}, \{4, 5\}\}, \{\{6, 7\}, \{8, 9\}\}\}}$

$\{\{\{\sqrt{2}, \sqrt{3}\}, \{2, \sqrt{5}\}\}, \{\{\sqrt{6}, \sqrt{7}\}, \{2\sqrt{2}, 3\}\}\}$

$\frac{1}{\{1, 2, 3, 4, 5\}}$

$\{1, \frac{1}{2}, \frac{1}{3}, \frac{1}{4}, \frac{1}{5}\}$

```
{1, 2, 3} + {4, 5, 6}
```

```
{5, 7, 9}
```

```
{1, 2, 3} {4, 5, 6}
```

```
{4, 10, 18}
```

$2^{\{0,1,2,3,4,5,6,7,8,9,10\}}$

```
{1, 2, 4, 8, 16, 32, 64, 128, 256, 512, 1024}
```

```
Cos[{0, π / 4, π / 2, 3 π / 4, π}]
```

$\left\{1, \frac{1}{\sqrt{2}}, 0, -\frac{1}{\sqrt{2}}, -1\right\}$

■ The Parts of a List

Here's a simple list:

```
alist = {2, x, y, {a, b}}
```

```
{2, x, y, {a, b}}
```

Notice that it has four parts.

```
Length[alist]
```

```
4
```

Here's the third part:

```
alist[[3]]
```

```
y
```

The fourth part is itself a list:

```
alist[[4]]
```

```
{a, b}
```

Here's the second part of that sublist:

```
alist[[4, 2]]
```

```
b
```

This gives the first and third parts:

```
alist[[{1, 3}]]
```

```
{2, y}
```

This gives the first *through* the third parts:

```
alist[[1 ;; 3]]
```

```
{2, x, y}
```

This uses `First` to extract the first part:

```
First[alist]
```

```
2
```

This uses `Last` to extract the last part:

```
Last[alist]
```

```
{a, b}
```

■ Creating Lists

Mathematica provides three functions that are especially useful for creating lists. These are `Range`, `Table` and `NestList`.

■ Range

`Range` can be used with one, two, or three arguments. With one argument, it returns a list of consecutive natural numbers beginning with 1.

```
Range[10]
```

```
{1, 2, 3, 4, 5, 6, 7, 8, 9, 10}
```

`Range[`a, b`]` returns a list containing $a, a+1, a+2, \ldots, a+n$, where $a+n \leq b < a+n+1$.

```
Range[4.5, 15.1]
```

```
{4.5, 5.5, 6.5, 7.5, 8.5, 9.5, 10.5, 11.5, 12.5, 13.5, 14.5}
```

A third argument specifies the increment. (The default is 1.)

$$\text{Range}\left[0,\ 1,\ \frac{1}{10}\right]$$

$$\left\{0,\ \frac{1}{10},\ \frac{1}{5},\ \frac{3}{10},\ \frac{2}{5},\ \frac{1}{2},\ \frac{3}{5},\ \frac{7}{10},\ \frac{4}{5},\ \frac{9}{10},\ 1\right\}$$

■ Table

`Table` provides an easy way of constructing many kinds of lists. The following computations illustrate its use.

$$\text{Table}\left[k^2,\ \{k,\ 10\}\right]$$

```
{1, 4, 9, 16, 25, 36, 49, 64, 81, 100}
```

```
Table[Table[i - j, {i, 4}], {j, 4}]
```

```
{{0, 1, 2, 3}, {-1, 0, 1, 2}, {-2, -1, 0, 1}, {-3, -2, -1, 0}}
```

```
% // MatrixForm
```

$$\begin{pmatrix} 0 & 1 & 2 & 3 \\ -1 & 0 & 1 & 2 \\ -2 & -1 & 0 & 1 \\ -3 & -2 & -1 & 0 \end{pmatrix}$$

■ NestList

`NestList` creates lists whose elements are terms of a *recursive sequence*; that is, given a function f and a starting point a_1, it creates a list containing $a_1, a_2, a_3, \ldots, a_n$, where $a_{k+1} = f(a_k)$. For example, ten terms of the arithmetic sequence defined by

$$a_1 = 3, \ a_{k+1} = 2a_k - 1, \ \ k = 1, 2, 3, \ldots$$

can be computed by first defining

```
f[x_] := 2 x - 1
```

and then entering

```
NestList[f, 3, 9]
{3, 5, 9, 17, 33, 65, 129, 257, 513, 1025}
```

Notice that `NestList` has three arguments. The first is the name of the function, the second is the first member of the list, and the third is the number of "steps" to be computed (which is one less than the length of the resulting list).

■ Manipulating Lists

■ Flatten

There are occasions when we need to simplify lists by "merging" smaller lists that it contains. The `Flatten` command does this. For example:

```
Flatten[{{1, 2, 3}, {4, 5}}]
{1, 2, 3, 4, 5}
Flatten[{{{x, y}, {3}}, {4, q}}]
{x, y, 3, 4, q}
```

■ Append and Prepend

We will often need to add elements to the end or beginning of an existing list. These tasks can be done with `Append` and `Prepend`. Here are two examples:

```
Append[{1, 2}, 3]
{1, 2, 3}
Prepend[{1, 2}, 0]
{0, 1, 2}
```

■ Union and Join

Merging two or more lists into one can be done with `Union` or `Join`. `Union` does *not* maintain the order of elements:

```
Union[{a, 2}, {v, 8}, {h, s}]
{2, 8, a, h, s, v}
```

but Join does:

```
Join[{a, 2}, {v, 8}, {h, s}]
{a, 2, v, 8, h, s}
```

Map and Apply

Map

A very useful method for applying a non-listable function to each element of a list is provided by Map. Suppose we have a list of ordered pairs of numbers such as

```
points = Table[{2 i, 5 - i}, {i, 8}]
{{2, 4}, {4, 3}, {6, 2}, {8, 1},
 {10, 0}, {12, -1}, {14, -2}, {16, -3}}
```

and we would like to create a list containing the sums of the numbers in each ordered pair in the list. To do this, we can create the function

```
addpairs[{x_, y_}] := x + y
```

and "map" it through the list of ordered pairs:

```
Map[addpairs, points]
{6, 7, 8, 9, 10, 11, 12, 13}
```

As an exercise, explain what goes on in the following:

```
Map[Flatten, {{3, {5, 6}}, {a, {b, c}}}]
{{3, 5, 6}, {a, b, c}}
```

Apply

Suppose that we have a function of two variables, say

```
vol[r_, h_] := π r^2 h
```

and that we would like to compute its value at a pair of numbers in a list, such as

```
measurements := {3.47, 5.12}
```

Now entering

```
vol[measurements]
vol[{3.47, 5.12}]
```

does not work. A very inconvenient, but effective, workaround is

```
vol[measurements[[1]], measurements[[2]]]
193.677
```

But a far simpler and more versatile approach is provided by Apply function:

```
Apply[vol, measurements]
193.677
```

It is usually easy to avoid such a situation (by defining `vol[{r_, h_}] := π r^2 h` in this case), but `Apply` does give us very nice way to compute the sum or product of the elements of a list:

```
Apply[Plus, {2, 5, 8, 12, 13}]
40
Apply[Times, {2, 5, 8, 12, 13}]
12 480
```

◆ Exercises

1. a) Use `Table` and `Prime` to generate a list of the first 100 prime numbers.
   b) Generate the same list using only `Prime` and `Range`.
2. Generate a list of values of the function $f(x) = \frac{\sin x}{x}$ for $x = .1, .2, \ldots, 1$, first using `Table`, then without `Table`.
3. Generate a list of the first 50 odd natural numbers, using:
   a) `Table`; b) `Range`; c) `NestList`
4. Generate a list of the first 21 powers of 2 (beginning with 2^0), using:
   a) `Table`; b) `Range`; c) `NestList`
5. Use `Table` to generate a list of ordered pairs $(x, f(x))$ for $x = 0, \frac{\pi}{12}, \frac{2\pi}{12} \ldots, \pi$, where $f(x) = \sin x$. Can you think of a way to do this without `Table`?
6. Create a list named `waves` that contains $\frac{1}{k} \sin kx$ for $k = 1, 2, 3, 4, 5$. Then plot the expressions in `waves` by entering

   ```
   Plot[Evaluate[waves], {x, 0, 2 π}]
   ```

   Now enter

   ```
   colors = Map[Hue, Range[.4, 1, .15]]
   Plot[Evaluate[waves], {x, 0, 2 π}, PlotStyle → colors]
   ```

   Then enter

   ```
   grays = Map[GrayLevel, Range[.8, 0, -.2]]
   Plot[Evaluate[waves], {x, 0, 2 π}, PlotStyle → grays]
   ```

7. a) The function

   ```
   f[x_] := x^5 - 2 x^2 - 3 x + 3
   ```

   has three real zeros. Plot its graph, and create a list named `guesses` that contains a rough estimate of each of the zeros.

   b) Define the function

   ```
   getZero[guess_] := FindRoot[f[x], {x, guess}]
   ```

   and find all three zeros of f with one command by entering

   ```
   Map[getZero, guesses]
   ```

1.5 Rules

Understanding rules is essential to making efficient use of *Mathematica*. For example, note that the `Solve` command returns its results as lists of rules:

```
soln = Solve[{x^2 + x + y^2 == 2, 2 x - y == 1}, {x, y}]
```

$$\left\{\left\{y \to \frac{1}{5}\left(-2-\sqrt{29}\right),\ x \to \frac{1}{10}\left(3-\sqrt{29}\right)\right\},\ \left\{y \to \frac{1}{5}\left(-2+\sqrt{29}\right),\ x \to \frac{1}{10}\left(3+\sqrt{29}\right)\right\}\right\}$$

To convert this answer to a list of pairs of numbers, we apply the rules to the list `{x,y}` as follows:

```
{x, y} /. soln
```

$$\left\{\left\{\frac{1}{10}\left(3-\sqrt{29}\right),\ \frac{1}{5}\left(-2-\sqrt{29}\right)\right\},\ \left\{\frac{1}{10}\left(3+\sqrt{29}\right),\ \frac{1}{5}\left(-2+\sqrt{29}\right)\right\}\right\}$$

Anticipating this in advance, we might have combined these steps by entering

```
{x, y} /. Solve[{x^2 + x + y^2 == 2, 2 x - y == 1}, {x, y}]
```

$$\left\{\left\{\frac{1}{10}\left(3-\sqrt{29}\right),\ \frac{1}{5}\left(-2-\sqrt{29}\right)\right\},\ \left\{\frac{1}{10}\left(3+\sqrt{29}\right),\ \frac{1}{5}\left(-2+\sqrt{29}\right)\right\}\right\}$$

The name of the *slash-dot* object "`/.`" that we use to apply rules is `ReplaceAll`. The following are some simple examples that illustrate its use.

```
Sqrt[x^2 - x + 1] /. x → 3
```

$\sqrt{7}$

```
x + y /. y → x
```

2 x

```
x y + y z + x z /. {x → a, y → b + c, z → 5}
```

5 a + 5 (b + c) + a (b + c)

◆ Exercises

1. Trigonometric identities provide a good context in which to learn about rules and gain a bit of insight into symbolic computation in general. For example, the sine addition formula can be applied via the rule

   ```
   sinAdd := Sin[x_ + y_] → Sin[x] Cos[y] + Cos[x] Sin[y]
   ```

   Notice what happens when the rule is applied to $\sin(3x + 5y)$:

   ```
   Sin[3 x + 5 y] /. sinAdd
   ```

   Cos[5 y] Sin[3 x] + Cos[3 x] Sin[5 y]

   The same rule provides the sine difference formula as well.

   ```
   Sin[t - ϕ] /. sinAdd
   ```

   Cos[ϕ] Sin[t] - Cos[t] Sin[ϕ]

This rule also handles expressions with three or more summands, provided we use `//.` (`ReplaceRepeated`) instead of `/.`:

```
Sin[a + b + c] //. sinAdd
Cos[b + c] Sin[a] + Cos[a] (Cos[c] Sin[b] + Cos[b] Sin[c])
```

a) Enter the definition of `sinAdd` and construct a similar rule, `cosAdd`, for the cosine addition formula. Test both rules on several different expressions.

b) Enter the following multiple-angle expansion formula for sine:

```
sinMult := Sin[n_Integer x_] → Sin[(n - 1) x + x] /. sinAdd
```

Check that this rule works properly by entering

```
{Sin[2 x], Sin[3 x]} //. sinMult
{2 Cos[x] Sin[x], 2 Cos[x]^2 Sin[x] + Cos[2 x] Sin[x]}
```

c) Construct a similar rule, `cosMult`, for the multiple-angle expansion formula for cosine. Test it on a few expressions.

d) Notice the result of repeatedly applying all four rules (followed by `Expand`) by entering

```
Sin[2 x + y] //. {sinAdd, cosAdd, sinMult, cosMult} // Expand
```

Then enter `Simplify[%]` to verify that the expansion is correct.

e) Define the function

```
trigExpand[expr_] :=
  expr //. {sinAdd, cosAdd, sinMult, cosMult} // Expand
```

and test it by entering

```
Cos[x + 2 y] + Sin[3 x - y] // trigExpand
% // Simplify
```

f) Finally, create a table of multiple angle formulas for sine by entering

```
Table[{Sin[k x], Sin[k x] // trigExpand}, {k, 1, 5}] // TableForm
```

and create a similar table of multiple angle formulas for cosine.

2. a) Use `NSolve` to find the zeros of the polynomial $f(x) = x^5 - 4x^4 + 12x^2 - 9x + 1$. Convert the result to a list of numbers.

b) Compute the sum of the zeros of f using `Total`. Combine the entire process into a single command.

c) Repeat the process in parts (a) and (b) after changing the coefficient of x^4 to 3, and then once again after changing the coefficient of x^4 to 1. Try changing the other coefficients to see if they affect the result. What do you conjecture about the sum of the zeros of a fifth-degree polynomial?

d) Compute the *product* of the zeros of f using `Apply` and `Times`. (See section 1.4.) Experiment with the coefficients to determine which affect the result. What do you conjecture about the product of the zeros of a fifth-degree polynomial?

e) Experiment with a few polynomials of other degrees. Do your conjectures depend on degree? Also, are your conjectures consistent with *linear* polynomials?

1.6 Graphics

■ Graphics Objects and Show

Graphics commands such as `Plot` create and display **graphics objects**.

```
graph1 = Plot[Sin[x] Cos[10 x], {x, 0, 2 π}]
```

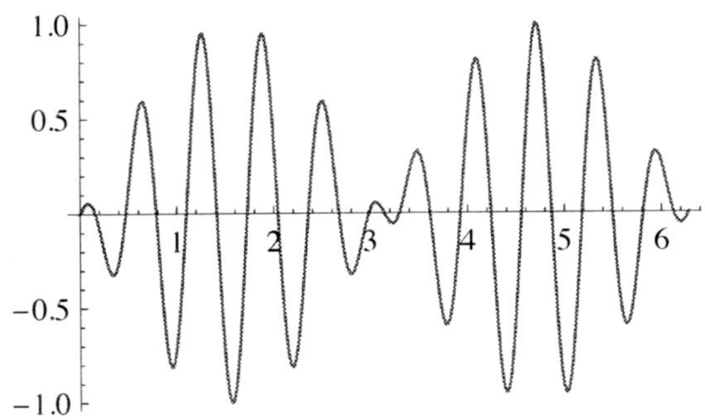

```
graph2 = Plot[{Sin[x], -Sin[x]}, {x, 0, 2 π}]
```

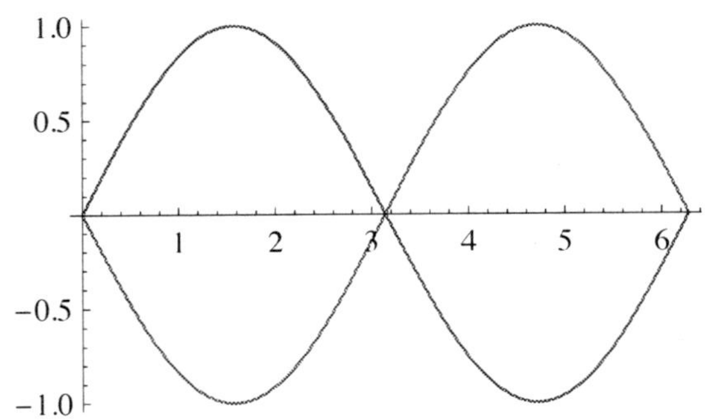

The `Show` command displays graphics objects, which may consist of two or more combined graphics objects.

```
Show[graph1, graph2]
```

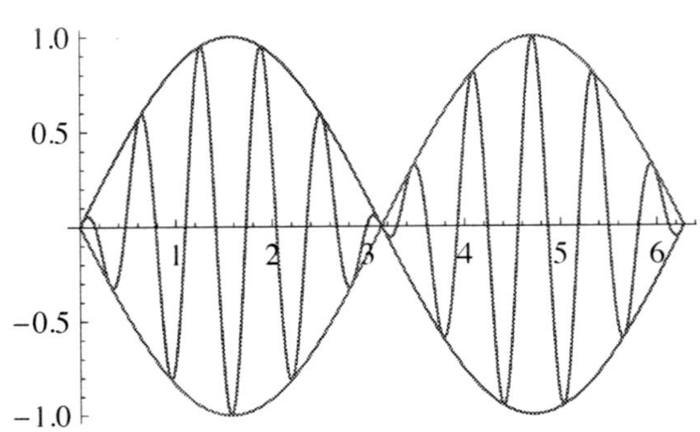

■ Graphics Primitives

Graphics primitives are the simple objects of which more complex graphics objects are built. Two- dimensional graphics primitives include `Point`, `Line`, `Circle`, `Disk`, `Rectangle`, `Polygon`, and `Text`.

The following defines a graphics primitive consisting of a series of line sergments connecting the specified points:

```
zigzag :=
 Line[{{1, 2}, {1, 1}, {3, 2}, {2, 2}, {1, 0}, {3, 1}, {2, 1}, {2, 0}}]
```

The `Graphics` command creates a graphics object from the graphics primitive.

```
Graphics[zigzag]
```

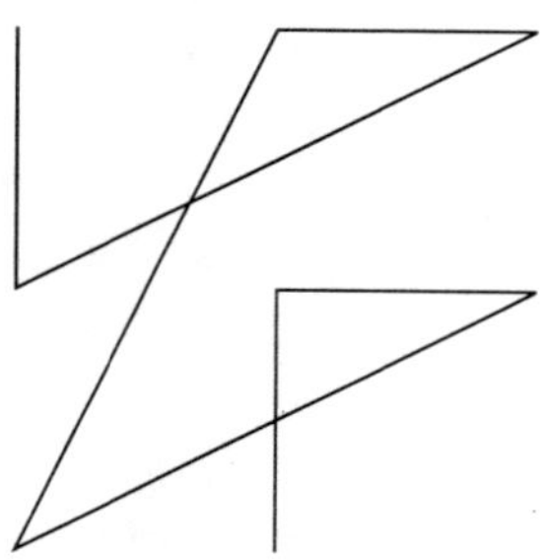

Here is a list of graphics primitives:

```
shapes = {Rectangle[{-2, 1}, {0, 2}], Circle[{1, 1}, 1],
   Disk[{0, 0}, .7], Text["rectangle", {-1.5, .8}],
   Text["disk", {-1, 0}], Text["circle", {1.5, 1.5}]};
```

This is the resulting graphics object:

```
Graphics[shapes, AspectRatio → Automatic]
```

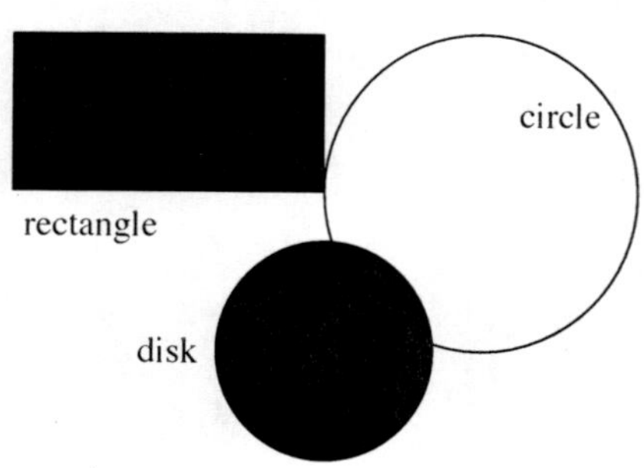

■ Graphics Directives

Graphics directives affect the way graphics primitives are displayed. Common graphics directives include `Opacity`, `PointSize`, and `Thickness`, as well as common colors such as `Red`, `Blue`, `Orange`, and so on. (`RGBColor`, `Hue`, `GrayLevel`, and `CMYKColor` are available for mixing your own colors.)

A graphics directive is associated with a graphics primitive by creating a list of the form {*directive*, *primitive*}. More than one primitive can be specified by creating a list of the form {*directive1*, *directive2*, ..., *primitive*}. The following suggests the many possibilities:

```
redRect = {Red, Opacity[.67], Rectangle[{-1, 1}, {1, 2}]};
thickCircle =
  {Thickness[.02], Opacity[.5], Green, Circle[{1, 1}, 1]};
purpleDisk = {Purple, Opacity[.4], Disk[{0, 1}, .7]};
Graphics[{redRect, thickCircle, purpleDisk}]
```

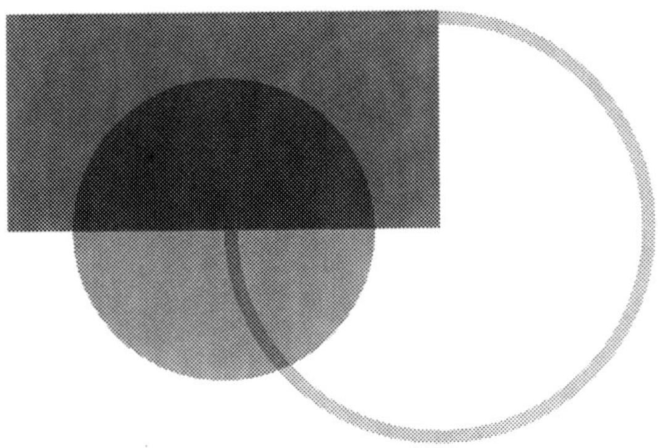

Often graphics directives are provided through *options* such as `PlotStyle`, `AxesStyle`, and `Background`. For example,

```
Plot[x^2, {x, -1, 1}, PlotStyle → {Thick, White},
 AxesStyle → Green, Background → GrayLevel[.5]]
```

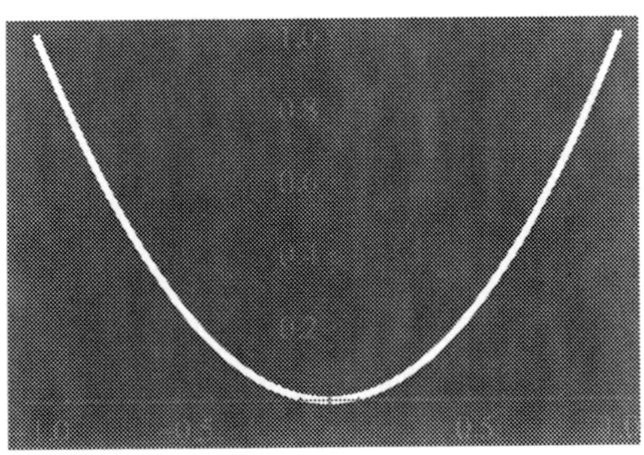

■ Suppressing and Combining Graphics

Suppose that we want to plot the parabola $y = x^2$ along with the circle of radius $1/2$ centered at $(0, 1/2)$. The following assigns names to plots of the parabola and the circle. *The output of each is surpressed by a semicolon*. (That's new in *Mathematica* 6.)

```
curve = Plot[x^2, {x, -1.4, 1.4}];
circ = Graphics[Circle[{0, .5}, .5]];
```

We can now combine the two graphics with `Show`.

```
Show[curve, circ, AspectRatio → Automatic]
```

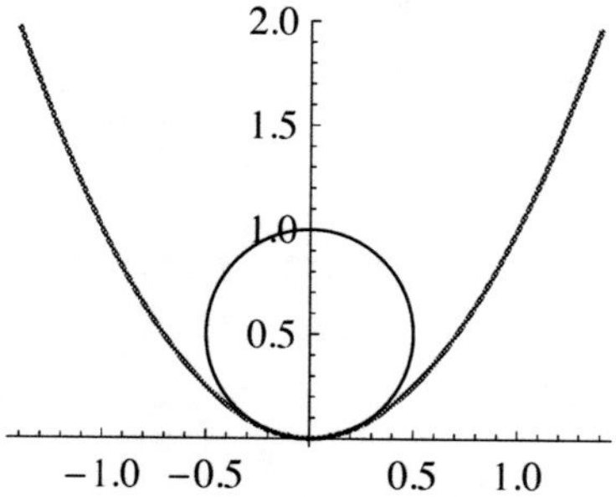

The same thing could be accomplished all at once by entering

```
Show[Plot[x^2, {x, -1.4, 1.4}],
 Graphics[Circle[{0, .5}, .5]], AspectRatio → Automatic]
```

(This behavior is new in *Mathematica* 6; in earlier versions the preceding command would have produced three separate plots.)

■ GraphicsRow and GraphicsGrid

Two very useful graphics commands are `GraphicsRow` and `GraphicsGrid`. With these we can create composite graphics objects containing rectangular arrays of individual graphics objects. For instance, let's create the following four graphics objects:

```
segment = Graphics[Line[{{0, 0}, {2, 2}}]];
circ = Graphics[Circle[{0, 0}, 1]];
parabola = Plot[x^2, {x, -1, 1}, Axes → None];
rect = Graphics[{Gray, Rectangle[{0, 0}, {1, 1}]}];
```

The following shows these four graphics objects in a one-by-four array:

```
GraphicsRow[{segment, circ, parabola, rect}]
```

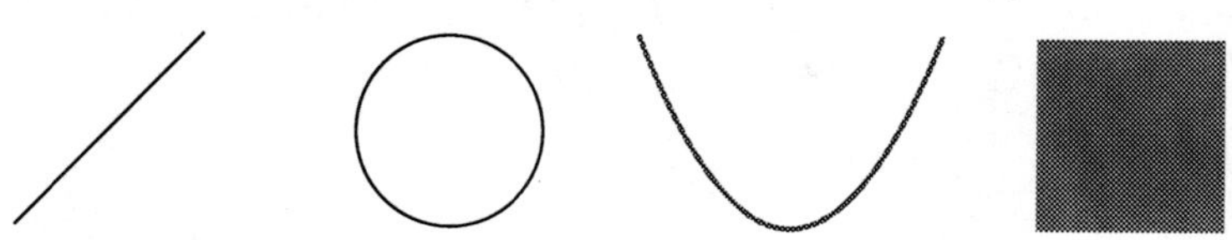

We would get a two-by-two array instead if we enter

```
GraphicsGrid[{{segment, circ}, {parabola, rect}}]
```

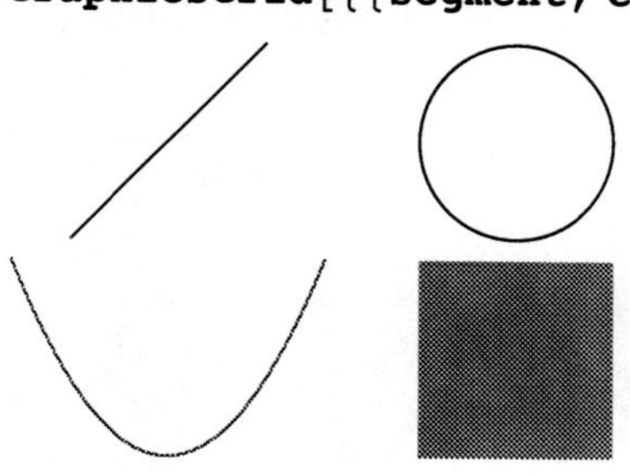

◆ Exercises

1. Predict the result of each of the following commands before entering it.

```
GraphicsRow[Table[Graphics[{Thickness[t], Orange, Circle[]}],
  {t, .02, .1, .02}]]

GraphicsRow[Table[
  Graphics[{PointSize[t], Point[{0, 0}]}], {t, .1, 1.1, .25}]]

GraphicsRow[Table[Graphics[{col, Disk[]}],
  {col, {Red, Blue, Green, Purple}}]]

GraphicsRow[Table[Graphics[{col, Disk[]}],
  {col, NestList[Lighter, Red, 3]}]]

GraphicsRow[
 Table[Graphics[{RGBColor[h, 0, 1 - h], Disk[]}], {h, 0, 1, .1}]]

Graphics[Table[{Hue[RandomReal[]],
   Disk[{Cos[t], Sin[t]}, .25]}, {t, π / 6, 2 π, π / 6}]]
```

1.7 Animate and Manipulate

One of the most instructive and fun features of *Mathematica* has always been its ability to animate graphics. New in *Mathematica* 6 is the function `Animate`, which provides a convenient, simple mechanism for creating animations.

For the sake of comparison, let's first create a simple table containing graphs of $y = \sin(x - \phi)$ for various values of the phase shift ϕ. Here ϕ will go from 0 to 2π in steps of 1 (by default).

```
Table[Plot[Sin[x - ϕ], {x, 0, 6 π},
  PlotRange → {{0, 6 π}, {-1.1, 1.1}}, AspectRatio → .2,
  Ticks → {Range[6] π, None}], {ϕ, 0, 2 π}]
```

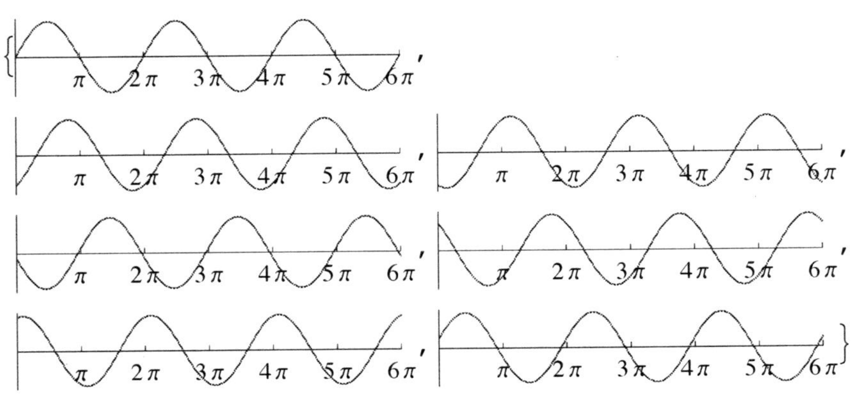

If we change `Table` to `Animate`, we get an animation instead, in which ϕ goes from 0 to 2π *continuously*. The controls allow you to start, stop, slow down, speed up, and reverse the animation. Moreover, the slider lets you "manually" move forward or backward through the animation.

```
Animate[Plot[Sin[x - ϕ], {x, 0, 6 π},
  PlotRange → {{0, 6 π}, {-1.1, 1.1}}, AspectRatio → .2,
  Ticks → {Range[6] π, None}], {ϕ, 0, 2 π}]
```

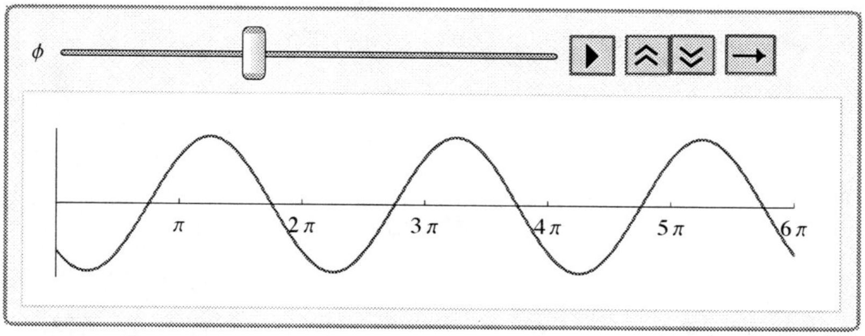

Using `Manipulate` instead, we get only the slider control. However, clicking on the ⊞ icon at the right end of the slider will reveal animation controls.

```
Manipulate[
 Plot[Sin[x - ϕ], {x, 0, 6 π}, PlotRange → {{0, 6 π}, {-1.1, 1.1}},
  AspectRatio → .2, Ticks → {Range[6] π, None}], {ϕ, 0, 2 π}]
```

💡 Notice in the preceding examples the specification of the `PlotRange` option. This is generally necessary to ensure that each plot created corresponds to the same rectangle in the plane, thereby producing an animation in which any fixed point remains still.

`Animate` and `Manipulate` can be used to animate or manipulate any type of expression, not just graphics.

```
Manipulate[Expand[(a + b)^n], {n, 1, 20, 1}] // boldEdge
```

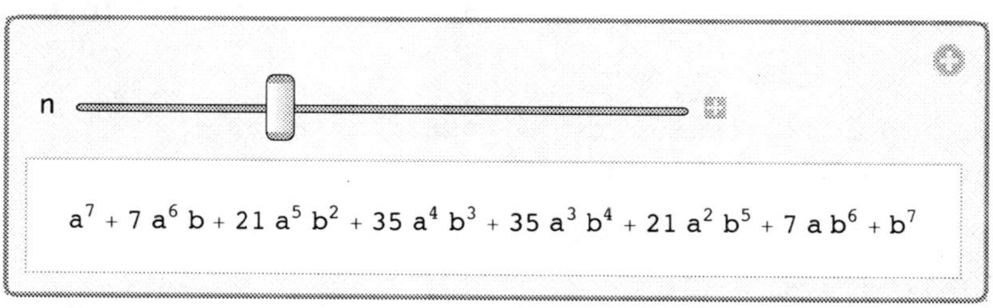

Here are two simple examples that use a different `ControlType`:

```
Manipulate[Expand[(a + b)^n],
 {n, 1, 15, 1}, ControlType → SetterBar]
```

```
Manipulate[x^y, {x, 1, 15, 1, ControlType → SetterBar},
 {y, 1, 15, 1, ControlType → SetterBar}]
```

◆ Exercises

1. The following creats a `Manipulate` object that plots the graph of $y = \sin(ax) + \sin(bx)$ for manipulable *angular frequencies* a and b of the individual terms. Enter the command and experiment with the two sliders. Then describe what you observe whenever the value of a "crosses over" the value of b or vice versa.

```
Manipulate[Plot[Sin[a t] + Sin[b t],
  {t, 0, 5}, PlotRange → {-2, 2}, PlotPoints → 40],
 {a, 50, 100}, {{b, 100}, 50, 100}]
```

This demonstrates the same phenomenon with sound:

```
Play[Sin[(950 + 10 t) t] + Sin[1000 t], {t, 0, 5}]
```

And this is the same thing in stereo:

```
Play[{Sin[(950 + 10 t) t], Sin[1000 t]}, {t, 0, 5}]
```

This is the phenomenon known as *beats*, which is useful in tuning string instruments by ear.

2. In this exercise, we will build up some simple and strangely interesting pieces of *Mathematica*-generated "art." Begin by entering

```
r := RandomReal[];
Show[Graphics[Line[{{r, r}, {r, r}}]], PlotRange → {{0, 1}, {0, 1}}]
```

This simply plots a random line segment within the square $-1 \le x \le 1, -1 \le y \le 1$. (Re-enter this a couple of times to observe the difference in the results.) Now enter the following several times. You should observe forty random segments each time.

```
Graphics[Table[Line[{{r, r}, {r, r}}], {40}],
 PlotRange → {{0, 1}, {0, 1}}]
```

Let's now give random color and thickness to the segments. Enter this a few times:

```
Graphics[Table[
  {RGBColor[r, r, r], Opacity[e^(-1.5 r)], Thickness[.01 + .02 r],
   Line[{{r, r}, {r, r}}]}, {40}], PlotRange → {{0, 1}, {0, 1}}]
```

Now create a composite graphic by entering

```
GraphicsRow[Table[Graphics[Table[{RGBColor[r, r, r],
    Thickness[.003 + .01 r], Line[{{r, r}, {r, r}}]}, {40}]], {5}]]
```

Enter the following to produce an animated piece on a black background:

```
Animate[Graphics[k; Table[
   {RGBColor[r, r, r], Opacity[.9 e^(-2 r)], Thickness[.01 + .02 r],
    Polygon[{{r, r}, {r, r}, {r, r}}]}, {30}],
  PlotRange → {{0, 1},  {0, 1}}, Background → Black], {k, 1, 20}]
```

Repeat the animation above, replacing the `Line[{{r,r},{r,r}}]` primitive with

a) `Circle[.25{1+2r,1+2r},7r/5]`

b) `Disk[{r,r},.5r]` c) `Polygon[{{r,r},{r,r},{r,r}}]`

1.8 Avoiding and Getting Out of Trouble

A Top Eleven List: Causes of *Mathematica* Problems

11. Forgetting that the natural log function is `Log`, not `Ln`

This is not peculiar to *Mathematica*; many advanced texts use this convention. However, if this bothers you, you can always define `Ln[x_]:=Log[x]`.

10. Typing an equation with one equal sign (=) instead of two (==)

A single equal sign is used only for assignments; an equation requires two. Recovering from this mistake often requires that you `Clear` a variable.

9. Forgetting to type a space between multiplied expressions

For example, if you accidentally type `xSin[x]`, *Mathematica* assumes that you are refering to a function named `xSin`.

8. Using parentheses instead of brackets or braces (or vice-versa)

Parentheses, brackets, and braces have very specific and different uses. Parentheses are used only for grouping within expressions, brackets enclose function arguments, and braces enclose members of a list.

7. Forgetting to load a package before referencing something in it

To correct the "shadowing" problem that results from this, you can enter `Remove[`*object*`]` where *object* is what you tried to use prior to loading the package.

6. Entering a command that relies on a previous definition that has not been entered during the current session

Whenever you resume work from a previous session, be sure that you re-enter commands in order from the top of your *Mathematica* notebook.

5. Doing an enormous symbolic computation instead of a simple numerical computation

See ■ Symbolic versus Numerical Computation below.

4. Making multiple definitions for one variable or function name

See ■ Multiple Definitions and Using the Question Mark below.

3. Spelling errors (including capitalization)

Enough said.

2. Forgetting to use a `Blank` when defining a function

See Section 1.3.

1. Forgetting to save your work before the inevitable crash

A word to the wise...

■ What to Do When You Run into Trouble

□ *Check for spelling mistakes, typos, and other syntax errors.*

□ *Look for online help.*

You can access the Documentation Center through the Help menu or by simply pressing your "help" key. If you want help on a particular command, option, etc., highlight that item before pressing "help."

□ *Clear variable names.*

Remember that entering `Clear[`*var1*, *var2*,...`]` clears variables.

□ *Clear everything.*

Here's a quick way to clear *all* previous definitions:

```
ClearAll["Global`*"]
```

□ *Quit and restart the kernel.*

Do this by choosing Quit Kernel: Local from your Kernel menu. You can then choose Start Kernel: Local from your Kernel menu or simply enter a command to start a new kernel session.

□ *Quit and restart Mathematica.*

Be sure to save your work first.

□ *Quit Mathematica and restart your computer.*

Again, be sure to save your work.

□ *Quit and restart your day. (Just kidding.)*

Seriously though, if you get frustrated, *take a break!*

💡 **Important note:** Although *Mathematica* remembers everything you enter during a particular session, it does not remember anything from a previous session or anything prior to clearing all variables or restarting the kernel. Since much of what you do in *Mathematica* depends on previously entered commands, you must be careful to reënter the commands that are needed after clearing all variables or restarting the kernel.

■ Interrupting Calculations

You will occasionally enter a command that takes *Mathematica* a very long time to evaluate. To stop a computation, select Abort Evaluation from the Evaluation menu. The keyboard shortcut for this is ⌘-. (command-period) on a Macintosh and CTRL-C on an Windows PC.

It is often necessary to press these keys repeated to interrupt a calculation, and sometimes there is no alternative but to quit the kernel.

Interpreting *Mathematica* Output When Things Don't Work

In many circumstances, *Mathematica* will give you a useful error message when a bad command is entered. Here are two examples:

```
Plot[x^2, {0, 1}]
```

Plot::pllim : Range specification {0, 1} is not of the form {x, xmin, xmax}. ≫

```
Plot[x^2, {0, 1}]
```

```
Solve[x^2 + 5 x = 2, x]
```

Set::write : Tag Plus in $5\,x + x^2$ is Protected. ≫
Solve::eqf : 2 is not a well-formed equation. ≫

```
Solve[2, x]
```

However, it is very common for *Mathematica* simply to give a problematic command back to you with no message. *Mathematica* does this whenever the syntax is correct, but no currently defined rules affect the result. For example,

```
aFunctionNotEntered[0]
```

```
aFunctionNotEntered[0]
```

and

```
Ln[1]
```

```
Ln[1]
```

When *Mathematica* simply gives a command back to you with no error message, it means that the syntax is okay, but something in the command is unrecognizable.

Symbolic Versus Numerical Computation

It is easy to run into major trouble by inadvertently asking *Mathematica* to create a huge symbolic expression. This is most likely to happen as a result of doing some kind of recursive calculation. For example, suppose we want to calculate several terms in the sequence defined by

$$x_0 = 1 \text{ and } x_{k+1} = 3 \sin x_k - x_k \text{ for } k = 0, 1, 2, \ldots$$

Here is a typical *Mathematica* approach: We'll define the function

```
f[x_] := 3 Sin[x] - x
```

and use `NestList` to compute terms in the sequence. This computes the first five terms:

```
NestList[f, 1, 4]
```

```
{1, -1 + 3 Sin[1], 1 - 3 Sin[1] - 3 Sin[1 - 3 Sin[1]], -1 + 3 Sin[1] +
  3 Sin[1 - 3 Sin[1]] + 3 Sin[1 - 3 Sin[1] - 3 Sin[1 - 3 Sin[1]]],
 1 - 3 Sin[1] - 3 Sin[1 - 3 Sin[1]] - 3 Sin[1 - 3 Sin[1] - 3 Sin[1 - 3 Sin[1]]] -
  3 Sin[1 - 3 Sin[1] - 3 Sin[1 - 3 Sin[1]] -
     3 Sin[1 - 3 Sin[1] - 3 Sin[1 - 3 Sin[1]]]]}
```

This is not exactly what we had in mind, is it? If we had asked for ten terms instead of five, the result would have filled several pages. (Try it.) If we had asked for thirty terms, and *if* the computation had *eventually* succeeded, the result would have contained more than 1/2 *billion* copies of the expression `Sin`! (Do yourself a favor; *don't* try it.)

So what should we do? We simply need to coerce *Mathematica* into doing the calculation numerically instead of symbolically, which is what we wanted to begin with! One simple way to do this is to start the sequence with the real number `1.` instead of the integer `1`. (Can you think of two other ways to accomplish the same thing?)

```
f[x_] := 3 Sin[x] - x; NestList[f, 1., 30]
```

```
{1., 1.52441, 1.47236, 1.51312, 1.48189, 1.50626,
 1.4875, 1.5021, 1.49082, 1.49959, 1.49281, 1.49807, 1.494,
 1.49716, 1.49471, 1.49661, 1.49514, 1.49628, 1.4954,
 1.49608, 1.49555, 1.49596, 1.49564, 1.49589, 1.4957,
 1.49585, 1.49573, 1.49582, 1.49575, 1.49581, 1.49576}
```

A useful tip: When attempting a complicated computation, *start small!* In other words, see what happens when you do three steps before you try to do thirty.

■ Multiple Definitions and Using the Question Mark

Suppose that we enter

f[x] = x^2 + 3 x

$3 x + x^2$

and we then realize that we forgot the `Blank` that we need to put beside the variable. So we then enter

f[x_] = x^2 + 3 x

$3 x + x^2$

and everything seems fine. *Later*... when working on a different problem, we redefine `f` as

```
f[x_] = x - 2
```

$-2 + x$

This function behaves as we expect; we find that its graph is the expected straight line, etc. But then we enter

g[x_] = f[x]^2

$(3 x + x^2)^2$

which does not give the function `g` that we expect. So what's going on here? *Mathematica* remembers our original, "erroneous" definition of the expression `f[x]`.

Using the Question Mark

To get information on any variable or other object, just type its name after a question mark. For example, to get information on f we'll enter

```
?f
```

```
Global`f

f[x] = 3 x + x^2

f[x_List] := 3 x

f[x_] := Which[x ≤ -1, 1, -1 < x ≤ 1, -x, x > 1, -1]

f[x_, y_] := x + y
```

This shows us that multiple definitions are associated with f. In fact, we could cause *Mathematica* to associate numerous other definitions with f:

```
f[x_, y_] := x + y
f[x_List] := 3 x
```

Now let's get information on f:

```
?f
```

```
Global`f

f[x] = 3 x + x^2

f[x_List] := 3 x

f[x_] := Which[x ≤ -1, 1, -1 < x ≤ 1, -x, x > 1, -1]

f[x_, y_] := x + y
```

When *Mathematica* encounters an expression involving f, it looks through the definitions associated with f until one makes sense for that expression. For example:

```
f[3, 5]
8
f[13]
11
f[{4, 7}]
{12, 21}
f[x]
3 x + x^2
```

As you may well imagine, this behavior of *Mathematica* can potentially be the source of all kinds of trouble.

The key to resolving difficulties caused by multiple definitions is to Clear the culprit variable. If you get into a really complicated mess, try quitting the kernel or entering

```
ClearAll["Global`*"]
```

The question mark is also useful for getting the "usage message" for built-in objects. Here are a few examples:

```
? Plot
```

Plot[f, {x, x_{min}, x_{max}}] generates a plot of f as a function of x from x_{min} to x_{max}.
Plot[{f_1, f_2, ...}, {x, x_{min}, x_{max}}] plots several functions f_i. >>

```
? $DisplayFunction
```

$DisplayFunction gives the default setting
for the option DisplayFunction in graphics functions. >>

```
? NestList
```

NestList[f, *expr*, n] gives a list of the results of applying f to *expr* 0 through n times.
>>

```
? /.
```

expr /. *rules* applies a rule or list of rules in an
attempt to transform each subpart of an expression *expr*. >>

■ Memory

Some of the most common difficulties that arise when using *Mathematica* are memory related—or rather, *lack-of*-memory related. *Mathematica* consists of two applications—the **kernel** and the **front end**—working together. Each of these has its own memory.

▪ Kernel Memory

The kernel is the part of *Mathematica* that does the computation. Many of the computations done by *Mathematica* involve highly complex algorithms and require a great deal of memory. In addition, the kernel remembers (by default) every command entered and every computation done in a given session. So it is easy to understand why running out of kernel memory—or experiencing poor performance due to use of virtual memory—can happen.

There are a couple of simple things that you can do to conserve memory:

- Set `$HistoryLength` to some small value such as 10 (*i.e.*, enter `$HistoryLength=10`). This causes the kernel to forget older input and output lines. The default value of `$HistoryLength` is `Infinity`.
- Use the `Share` command occasionally:

```
Share[]

1 010 880
```

This causes stored expressions to "share" subexpressions, thus reducing the amount of memory used. The output shows the number of freed bytes.

Also, see ■ Symbolic versus Numerical Computation in the preceeding section.

■ Front End Memory and File Size

Front-end memory and notebook file size usually only become an issue when your notebook contains a lot of graphics. While the computations that create a graphic are done by the kernel, the code that actually produces the graphic is stored in the front end's memory.

By deleting graphics cells—especially cells containing three-dimensional graphics or graphics created with a high value for the `PlotPoints` option—you can greatly decrease the amount of front-end memory used.

When you save your work, it is information in the front end's memory that you are saving—in the form of a *Mathematica* "notebook." When a notebook becomes very large, you can usually remedy the situation by deleting graphics cells before saving. Graphics can always be reproduced, as long as the commands are saved.

A handy feature is the Delete All Output item in the Cell menu. This will let you quickly save the essence of your work in a very small file.

■ Exercises

1. Purposely commit each of the errors described in the eleven causes of problems outlined above—with the exception of numbers 4 and 1. In cases where no consequence is immediately evident, construct a subsequent scenario that exposes the error.

1.9 Turning a Notebook into a Report

You will likely be asked to put the work you do in *Mathematica* into a form that will be presentable enough to submit to your professor. Fortunately, the *Mathematica* front end serves as a very versatile word processor. In fact, this entire manual was written with *Mathematica*.

■ Cell Styles

You should always provide comments and narrative along with your calculations (whether you're using *Mathematica* or pencil and calculator). Any cell that contains text should be given Text Style by selecting Text from the Style submenu of the Format menu before you begin typing text into the cell. You can also give an existing cell Text Style by highlighting the cell bracket and selecting Text from the same menu.

You should also use Title, Section, Subsection cells, etc., to organize your notebook. These items are also in the Style submenu of the Format menu.

Aside from resulting in much nicer looking work, this is also important because when an Input cell contains text, all kinds of errors and garbage can result if it is accidentally evaluated.

```
For example notice what happens when I enter this
```

```
i enter example For happens notice this what when
```

```
or when I enter this.
```

Syntax::tsntxi : "this." is incomplete; more input is needed.

Syntax::tsntxi : "this." is incomplete; more input is needed.

Syntax::sntxi : Incomplete expression; more input is needed.

Moreover, if Input cells only contain valid *Mathematica* commands, it is possible to evalute all of them in order by selecting Evaluate Notebook from the Evaluation menu without making a big mess of your the notebook.

■ Page Breaks

Bad page breaks usually involve a large amount of blank space at the bottom of a page. Frequently this is caused by a graphic being just a little too large to fit on the current page. The default size of *Mathematica* graphics is larger than it usually needs to be for printing. So by resizing (*i.e.*, shrinking) graphics, you can avoid a lot of bad page breaks. A graphic can be resized by clicking on it and dragging a corner. Also, a good way to get smaller, consistently-sized graphics from `Plot`, for instance, is to use `SetOptions`:

```
SetOptions[Plot, ImageSize → 200]
```

How can you tell where page breaks will occur before you print? You can select Show Page Breaks from the Printing Settings submenu of the File menu. (In *Mathematica* 5.x, this is located in the Format menu.)

You can also *force* a page break between two cells by selecting Page Break from the Insert Menu.

■ Other Tips

In the Printing Settings submenu of the File menu, you'll see Printing Options... In the resulting dialog box, you can set margins and specify whether or not to print cell brackets.

In the Style Sheet submenu of the Format menu, you can choose from among several standard style sheets. The choice of style sheet affects the appearance of title and section cells, background color, etc. Experiment to find one that you like. But don't be surprised if you end up preferring the default.

◆ Exercises

1. Write a short but detailed report (2-3 pages) in *Mathematica* on any one of the following topics. Your report must include input, output, text, section, and title cells.

   a) The rational root theorem for polynomials.

   b) How to find the inverse of a one-to-one function.

   c) Even functions and odd functions.

   d) The unit circle and the graphs of $\sin x$ and $\cos x$.

   e) The compound interest formula.

   f) Rational functions with slant asymptotes.

1.10 Miscellaneous Advice

■ The Cube Root Function

When you ask *Mathematica* for the cube root (or any odd root) of a negative number, it returns a complex number. This complex number is indeed the *principal value* the cube root function defined on the complex numbers.

```
∛-8.
```

```
1. + 1.73205 i
```

However, this is not what we want when we talk about the cube root function defined on the real numbers. A simple remedy is to define your own cube root function as follows:

```
Cbrt[x_] := Sign[x] ∛Abs[x]
Plot[Cbrt[x], {x, -4, 4}]
```

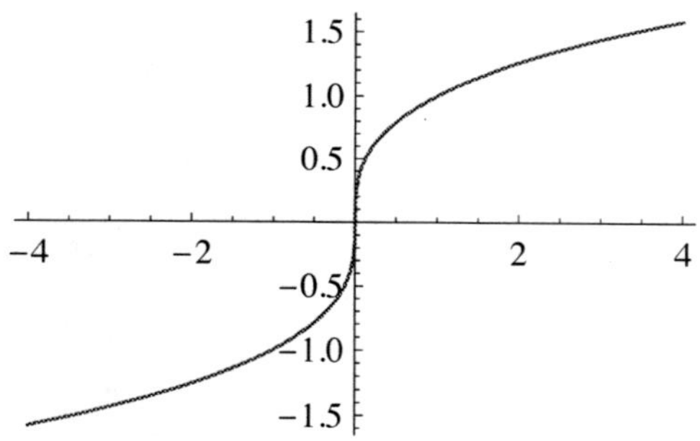

■ Custom Initialization

You may find that there are certain commands you want to enter every time you run *Mathematica*. For instance, if you prefer that the curves drawn by `Plot` are always thicker than the default thickness, you can avoid specifying that with the `PlotStyle` option every time you use plot by entering

```
SetOptions[Plot, PlotStyle → Thickness[.005]]
```

But that will only be in effect during the current kernel session; that is, the next time you start *Mathematica*, you'll have to enter that command again.

Assuming that you have appropriate priviledges on the computer on which you're working, you can execute a group of commands automatically each time the kernel starts up by placing those commands in a file named init.m that's located in a particular directory. An easy way to find and open that file is to enter

```
ToFileName[{$UserBaseDirectory, "Kernel"}, "init.m"]
NotebookOpen[%]
```

You can type in the commands you want executed automatically upon startup and then save and close the file. Actually, it's best to type the commands in an open notebook, make sure they work correctly, and then copy and paste them into the init.m file.

Here are a few examples of things you might want to put in init.m:

- This sets options for `Plot` so that it draws thicker curves and makes plots 3 inches wide.

```
SetOptions[Plot,
  {PlotStyle → Thickness[Medium], ImageSize → 3 * 72}];
```

- This defines a real cube root function.

```
Cbrt[x_] := Sign[x] ∛Abs[x] ;
```

- This sets `$HistoryLength` to conserve memory.

```
$HistoryLength = 10
```

2 Limits and the Derivative

The graphical, algebraic, and numerical capabilities of *Mathematica* allow us to explore the basic concepts behind limits and derivatives. In this chapter we'll explore some of these capabilities and concepts.

2.1 Limits

The notion of limit is the basis of all of calculus. From one point of view, we can think of a limit as a means of describing a function's behavior near a point where the function cannot be evaluated.

▪ Example 2.1.1

The function

```
f[x_] := Sin[x]/x
```

is undefined at $x = 0$, but defined at all other x. The question arises, then, as to how the function behaves for x nearby, but not equal to, 0. More precisely, what happens to the values of the function as x *approaches* 0? Let's first investigate this by making a few of tables of values. The following creates three. (Notice the use of `Quiet` to surpress division-by-zero error messages.)

```
Grid[{{TableForm@Table[{x, f[x]}, {x, -.4, .4, .1}],
    TableForm@Table[{x, f[x]}, {x, -.04, .04, .01}],
    TableForm@Table[{x, f[x]}, {x, -.004, .004, .001}]
   }}, Spacings → 2] // Quiet
```

```
-0.4 0.973546        -0.04 0.999733        -0.004 0.999997
-0.3 0.985067        -0.03 0.99985         -0.003 0.999999
-0.2 0.993347        -0.02 0.999933        -0.002 0.999999
-0.1 0.998334        -0.01 0.999983        -0.001 1.
0.   Indeterminate   0.    Indeterminate   0.     Indeterminate
0.1  0.998334        0.01  0.999983        0.001  1.
0.2  0.993347        0.02  0.999933        0.002  0.999999
0.3  0.985067        0.03  0.99985         0.003  0.999999
0.4  0.973546        0.04  0.999733        0.004  0.999997
```

The values in these tables indicate that $f(x)$ becomes very close to 1 as x approaches 0. This observation is confirmed by the graph of the function.

```
Plot[f[x], {x, -2 π, 2 π}]
```

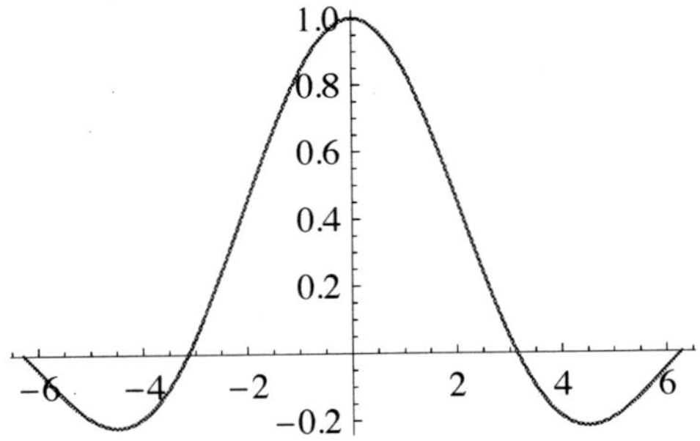

So we see that even though $f(0)$ is undefined, values of x near 0 produce values of $f(x)$ near 1. Moreover, it appears that $f(x)$ can be made to lie as close as we like to 1 by taking x sufficiently close to 0. We describe this by saying that 1 is the **limit** of $f(x)$ as x approaches 0 and express it mathematically by writing

$$\lim_{x\to 0} f(x) = 1.$$

We can also use *Mathematica*'s `Limit` function to calculate this limit:

```
Limit[f[x], x → 0]
```

```
1
```

▪ Example 2.1.2

Consider the function

$$f(x) = \left(x^3 - 1\right)^{4/3} (x - 1)^{-1},$$

which is undefined at $x = 1$ but defined everywhere else. Let's investigate the behavior of this function near $x = 1$. Again, we'll start by examining tables of values. But because *Mathematica*'s cube root function returns complex cube roots of negative numbers, we need to be careful how we enter the definition of f. We can avoid this difficulty by using the fact that $x^{4/3} = \left(x^4\right)^{1/3}$.

```
f[x_] := ((x^3 - 1)^4)^(1/3) (x - 1)^-1
```

```
Grid[{{TableForm@Table[{x, f[x]}, {x, .97, 1.03, .01}],
    TableForm@Table[{x, f[x]}, {x, .997, 1.003, .001}],
    TableForm@Table[{x, f[x]}, {x, .9997, 1.0003, .0001}]
  }}, Spacings → 2] // Quiet
```

```
0.97 -1.29145         0.997 -0.621533         0.9997 -0.289531
0.98 -1.14345         0.998 -0.543684         0.9998 -0.252962
0.99 -0.919803        0.999 -0.432098         0.9999 -0.200803
1.   Indeterminate    1.    Indeterminate     1.     Indeterminate
1.01 0.944661         1.001 0.433252          1.0001 0.200857
1.02 1.20609          1.002 0.546591          1.0002 0.253097
1.03 1.39901          1.003 0.626525          1.0003 0.289763
```

These tables do not make it obvious what $f(x)$ does as x approaches 1, although the approximate symmetry about $x = 1$ suggests that the limit is 0. This is confirmed by the graph of f. (Notice that the graph indicates that x must be *very* close to 1 for $f(x)$ to be at all close to 0.)

```
Plot[f[x], {x, .5, 1.4}]
```

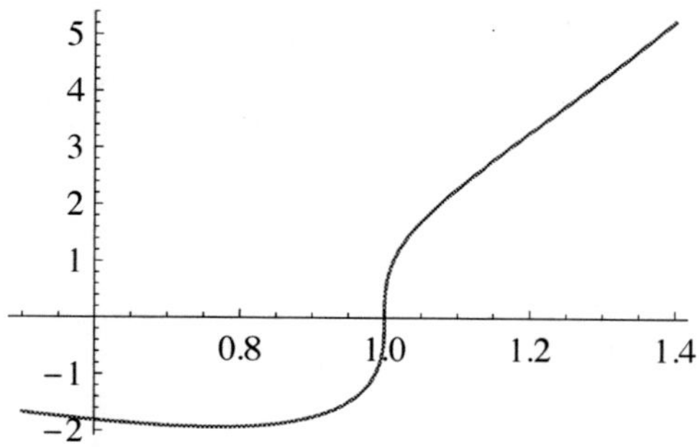

Thus we see that $\lim_{x \to 1} f(x) = 0$, which is further confirmed by `Limit`:

```
Limit[f[x], x → 1]
0
```

It is important to note here that we have have not proven either of our stated limits in the previous examples. Nor have we even given a truely meaningful definition of we mean by the limit of a function f as x approaches a number a; we have merely relied on an intuitive notion. See your primary textbook for both an intuitive definition and the precise, technical definition.

■ One-sided Limits

Limit often fail to exist. Sometimes when a limit does not exist, one-sided (i.e., left- or right-sided) limits do exist. *Mathematica*'s `Limit` function finds left- or right-sided limits, when we supply the `Direction` option. `Direction→1` results in the left-sided limit, and `Direction→-1` results in the right-sided limit. (These settings make sense if we think of the left-sided limit as the limit in the positive direction and the right-sided limit as the limit in the negative direction.)

▪ Example 2.1.3

Consider the function

```
f[x_] := (2 x + Abs[x])/Abs[x] Cos[x]
```

which is undefined at $x = 0$. To compute $\lim_{x \to 0^-} f(x)$, the left-sided limit as x approaches zero, we'll enter

```
Limit[f[x], x → 0, Direction → 1]
-1
```

For $\lim_{x\to 0^+} f(x)$, the right-sided limit as x approaches zero, we'll enter

```
Limit[f[x], x → 0, Direction → -1]
```

```
3
```

The meaning of these one-sided limits is indicated in the graph of the function:

```
Plot[f[x], {x, -5, 5}]
```

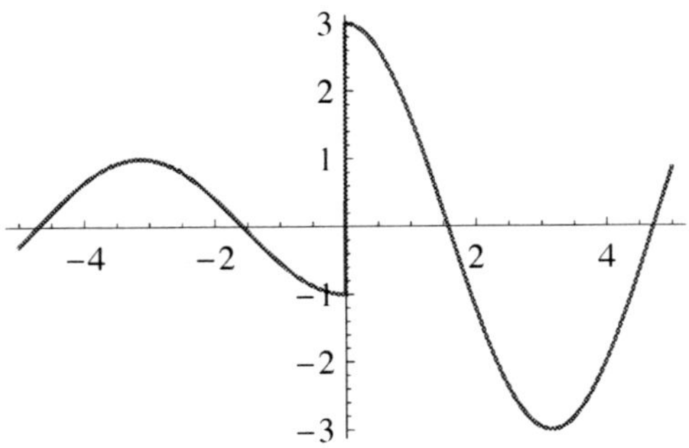

Since $\lim_{x\to 0^-} f(x) \neq \lim_{x\to 0^+} f(x)$, we conclude that $\lim_{x\to 0} f(x)$ does not exist. This function has what is sometimes called a "jump discontinuity" at $x = 0$ by virtue of the fact that both one-sided limits exist but with different values.

⚠ **Warning:** By default, *Mathematica*'s `Limit` function returns a right-sided limit. For example, with the function defined above, we see

```
Limit[f[x], x → 0]
```

```
3
```

Technically this is incorrect, since $\lim_{x\to 0} f(x)$ does not exist. This issue gives us a reason to create our own limit command. The following command returns the limit if the left- and right-sided limits agree and returns a "doesn't exist" message otherwise.

```
theLimit[expr_, xtoa_] :=
 Module[{rlim = Limit[expr, xtoa, Direction → -1]}, If[
   rlim == Limit[expr, xtoa, Direction → 1], rlim, "doesn't exist"]]
```

Now we find correctly that

```
theLimit[f[x], x → 0]
```

```
doesn't exist
```

■ Infinite Limits and Vertical Asymptotes

Sometimes one-sided limits fail to exist because the graph of the function in question has a vertical asymptote. In such cases, one-sided limits can be assigned a value of either ∞ or $-\infty$ to described the manner in which the graph approaches the vertical asymptote.

■ Example 2.1.4

Each of the functions

$$f[x\_] := \frac{1}{x-1};\quad g[x\_] := \frac{1}{(x-1)^2}$$

has a vertical asymptote at $x = 1$. One-sided limits describe the behavior of each function at the vertical asymptote. Notice that $f(x)$ is negative for $x < 1$ and positive for $x > 1$.

```
{Limit[f[x], x → 1, Direction → 1],
 Limit[f[x], x → 1, Direction → -1]}
```

```
{-∞, ∞}
```

```
Plot[f[x], {x, -2, 4},
     PlotRange → {-10, 10}]
```

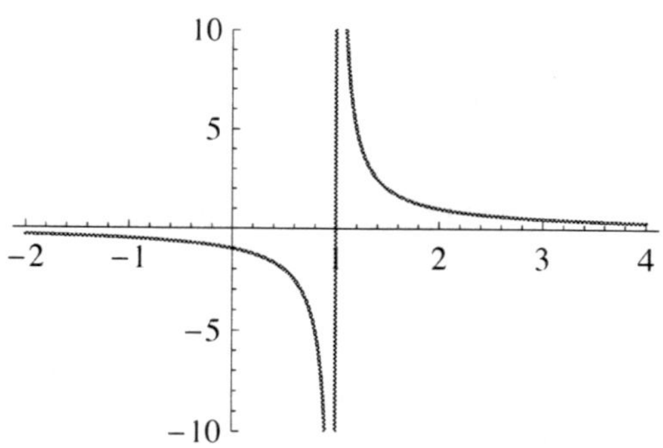

Notice that $g(x)$ is positive for $x < 1$ and for $x > 1$; so each one-sided limit as $x \to 1$ is $+\infty$.

```
{Limit[g[x], x → 1, Direction → 1],
 Limit[g[x], x → 1, Direction → -1]}
```

```
{∞, ∞}
```

```
Plot[g[x], {x, -2, 4},
   PlotRange → {0, 30}]
```

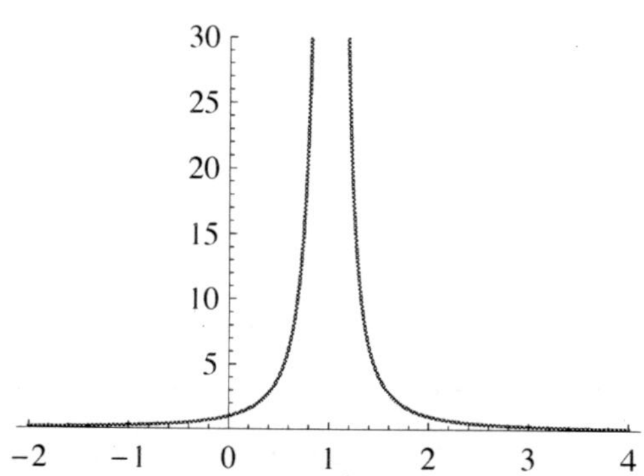

■ Limits at $\pm\infty$ and Horizontal Asymptotes

While infinite limits occur at vertical asymptotes of a function, limits *at* $\pm\infty$ determine horizontal asymptotes. The limit of $f(x)$ as $x \to \infty$ is a number that $f(x)$ approaches as x increases without bound. The limit of $f(x)$ as $x \to -\infty$ is a number that $f(x)$ approaches as x decreases without bound.

■ Example 2.1.5

The function

```
f[x_] := ArcTan[x]
```

has distinct limits at $\pm\infty$, as evidenced below.

```
Plot[f[x], {x, -20, 20}]
```

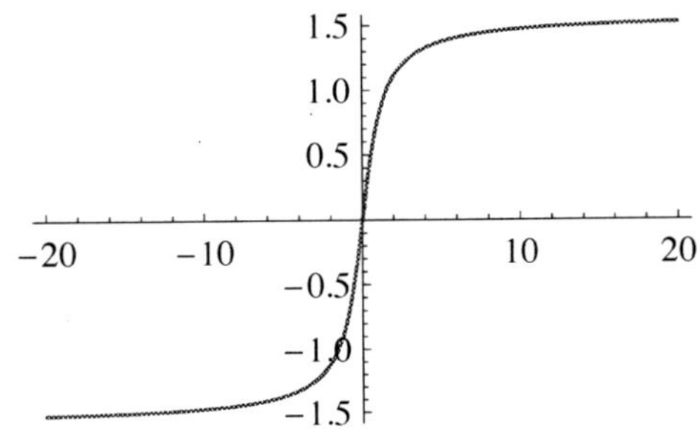


```
{Limit[f[x], x → ∞], Limit[f[x], x → -∞]}
```

$\left\{\frac{\pi}{2}, -\frac{\pi}{2}\right\}$

■ Example 2.1.6

The function

```
f[x_] := e^(-x/10) Cos[x^2]
```

has a limit at ∞, but no limit at $-\infty$.

```
Plot[f[x], {x, -12, 24},
        PlotRange → All]
```

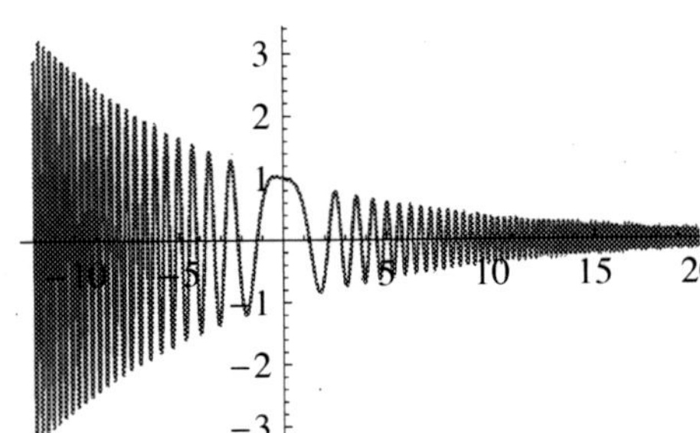


```
{Limit[f[x], x → -∞], Limit[f[x], x → ∞]}
{Interval[{-∞, ∞}], 0}
```

Notice here how *Mathematica* describes the behavior of $f(x)$ as $x \to -\infty$, indicating that every real number is a value of f arbitrarily often as $x \to -\infty$.

▪ Example 2.1.7

The rational function

```
f[x_] := (2 x^2 + 5 x - 1)/(x^2 + 1)
```

$$\texttt{f[x\_]} := \frac{2\,x^2 + 5\,x - 1}{x^2 + 1}$$

has the same limit at both $\pm\infty$.

```
{Limit[f[x], x → -∞], Limit[f[x], x → ∞]}
```

```
{2, 2}
```

```
Plot[{f[x], 2}, {x, -20, 20}, PlotRange → All]
```

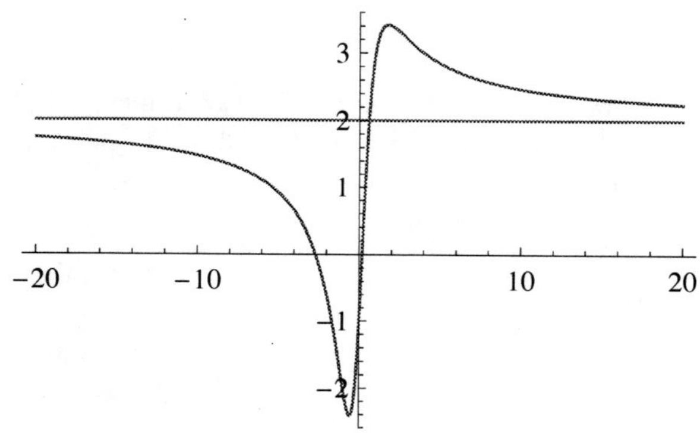

▪ Limits of Expressions with Symbolic Parameters

As we will see in the next section, limits such as

$$\lim_{h\to 0} \frac{\sqrt{4+h} - 2}{h}$$

are very important in calculus. This particular limit is easily found to be 1/4.

$$\texttt{Limit}\left[\frac{\sqrt{4+h} - \sqrt{4}}{h},\ h \to 0\right]$$

$$\frac{1}{4}$$

Notice, however, that this limit is just the value at $x = 4$ of the function f defined by

$$\texttt{f[x\_]} = \texttt{Limit}\left[\frac{\sqrt{x+h} - \sqrt{x}}{h},\ h \to 0\right]$$

$$\frac{1}{2\sqrt{x}}$$

The point is that even though the function f is defined in terms of a limit, it ends up having a simple "closed form."

The following limit is typical of an important kind of limit we will encounter in Chapter 4.

```
Limit[Sum[Cos[k π/(3 n)] π/(3 n), {k, 1, n}], n → ∞]
```

$$\frac{\sqrt{3}}{2}$$

Notice that this limit is the value at $x = \pi/3$ of the function f defined by

```
f[x_] = Limit[Sum[Cos[k x/n] x/n, {k, 1, n}], n → ∞]
Sin[x]
```

◆ Exercises

In Exercises 1-12, estimate the requested limit by computing tables of values and plotting the graph. If possible, make a conjecture about the exact value of the limit. Then use `Limit` or `theLimit` to confirm your conjecture.

1. $\lim_{x\to 1} \frac{x^3+2x-3}{x-1}$

2. $\lim_{x\to 1} \frac{\sqrt{2x^3-x^2-4x+3}}{x-1}$

3. $\lim_{x\to 0} x\sin(1/x)$

4. $\lim_{x\to 0} \frac{\sin\left(e^{2x}-1\right)}{x}$

5. $\lim_{x\to 0} \frac{\sin 5x}{x^2}$

6. $\lim_{x\to 0^-} \frac{|\sin x|}{x}$

7. $\lim_{x\to 0^+} \frac{|x|-x^2}{x}$

8. $\lim_{x\to 1^-} \frac{x^3+2x-4}{x-1}$

9. $\lim_{x\to 0^+} \frac{x^3+2x-3}{x}$

10. $\lim_{x\to \infty} \frac{\sqrt{3x^2+1}}{2x-5}$

11. $\lim_{x\to \pi/4} \frac{1-\tan x}{\sin x-\cos x}$

12. $\lim_{x\to 0^+} x\ln x$

In Exercises 13-15, plot the given function; then find both the left- and right-sided limits at each specified point. The *Mathematica* definition for the function in Exercise 13 is given.

13. $f(x) = \begin{cases} -x-2, & \text{if } x < -1 \\ -1, & \text{if } -1 \le x \le 0 \\ \sin(2\pi x), & \text{if } 0 < x < 1 \\ x, & \text{if } 1 \le x \end{cases}$ at $x = -1,\ 0,\ 1$

```
f[x_] = Piecewise[{{-x - 2, x < -1},
    {-1, -1 ≤ x ≤ 0}, {Sin[2 π x], 0 < x < 1}, {x, 1 ≤ x}}]
```

14. $f(x) = \begin{cases} 2, & \text{if } x < 0 \\ 1, & \text{if } 0 \le x < 2 \\ 3, & \text{if } 2 \le x \end{cases}$

at $x = 0,\ 2$

15. $f(x) = \begin{cases} 1, & \text{if } x < -1 \\ |x|, & \text{if } -1 \le x < 1 \\ 1, & \text{if } 1 \le x \end{cases}$

at $x = -1,\ 1$

16. Find: a) $\lim_{h\to 0} \frac{(x+h)\,e^{x+h}-x\,e^x}{h}$ b) $\lim_{x\to 1} \frac{(a\,x-1)^2-(a-1)^2}{x-1}$

2.2 The Derivative

The fundamental geometric problem with which we are concerned is that of finding the slope of the *tangent line* to the graph of a function f at a given point $(a, f(a))$. Slopes of lines are easy to compute by means of the slope formula

$$m = \frac{y_2 - y_1}{x_2 - x_1},$$

provided we know two points on the line. The difficulty that arises in finding the slope of the tangent line to a graph is that only one point on the line is immediately known, namely $(a, f(a))$. So we take an approach that is the essence of calculus: *We approximate the quantity and take the limit as we refine our approximations.*

The slope of the tangent line at $(a, f(a))$ can be approximated by the slope of the *secant line* through $(a, f(a))$ and a nearby point $(a+h, f(a+h))$ on the graph, where h is some small number. The slope of such a secant line is easily calculated as

$$m_a(h) = \frac{f(a+h) - f(a)}{h}$$

The equation of the secant line is then

$$y = f(a) + m_a(h)\,(x - a).$$

▪ Example 2.2.1a

Let's graph the function

```
f[x_] := x (2 - x)
```

along with the secant line through $(.5, f(.5))$ and $(1, f(1))$. We'll first define

```
secLine[a_, h_][x_] := f[a] + (f[a+h]-f[a])/h (x - a)
```

The desired secant line is the graph of

```
q[x_] = secLine[.5, .5][x]
0.75 + 0.5 (-0.5 + x)
```

Now we're ready to plot the graph.

```
Plot[{f[x], q[x]}, {x, 0, 1.5}]
```

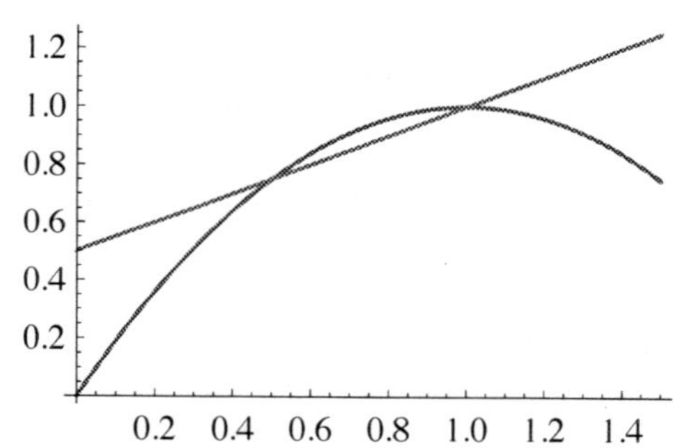

■ The Slope of the Tangent Line

The slope of the tangent line at $(a, f(a))$ is given by

$$m_{\tan} = \lim_{h \to 0} m_a(h) = \lim_{h \to 0} \frac{f(a+h) - f(a)}{h}$$

We can illustrate this very important limit process by graphing secant lines for a few decreasing values of h.

■ Example 2.2.1b

Again we'll illustrate with $f(x) = x(2 - x)$ and $a = .5$, plotting secant lines corresponding to $h = .5, .3, .15,$ and $.01$.

```
Plot[{f[x], secLine[.5, .5][x], secLine[.5, .3][x],
  secLine[.5, .15][x], secLine[.5, .01][x]}, {x, 0, 1.5}]
```

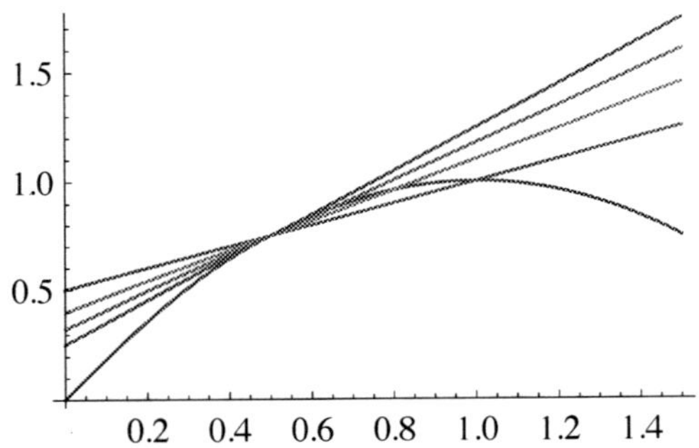

The secant line with $h = .01$ provides a good approximation to the tangent line in this case. This is seen in the following plot.

```
Plot[{f[x], secLine[.5, .01][x]}, {x, 0, 1.5}]
```

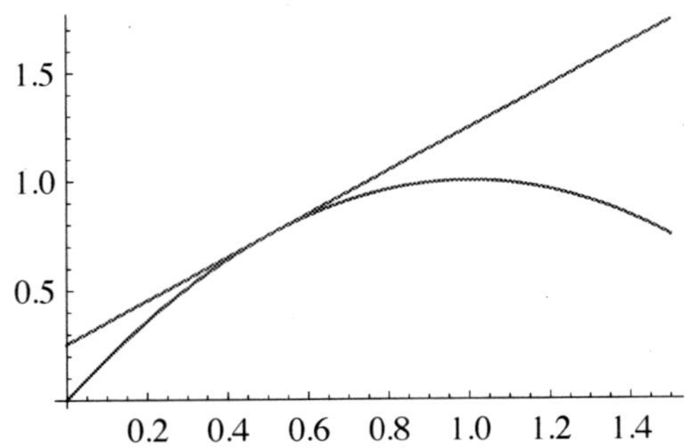

■ A Better Slope Approximation

As indicated above, the slope of the tangent line to the graph of $y = f(x)$ at $(a, f(a))$ may be approximated with the *forward difference approximation*:

$$m_{\tan} \approx \frac{f(a+h) - f(a)}{h} \quad \text{for small } h.$$

It turns out that a far better approximation, for a given h, is provided by the *central difference approximation*,

$$m_{\tan} \approx \frac{f(a+h) - f(a-h)}{2h} \quad \text{for small } h,$$

in which the slope of the tangent line at $(a, f(a))$ is approximated by the slope of the line through $(a-h, f(a-h))$ and $(a+h, f(a+h))$.

■ Example 2.2.2

To illustrate the superiority of the central difference approximation to the forward difference approximation, let's consider the function

```
f[x_] := √x (4 - x)
```

Forward and central difference approximations can be defined as

```
mfwd[a_, h_] := (f[a + h] - f[a]) / h;
mcntrl[a_, h_] := (f[a + h] - f[a - h]) / (2 h)
```

Now let's find functions that define secant lines passing through $(1, f(1))$ and with $h = .5$, one with slope given by the forward difference approximation and the other with slope given by the central difference approximation.

```
secfwd[x_] = f[1] + mfwd[1, .5] (x - 1)
seccntrl[x_] = f[1] + mcntrl[1, .5] (x - 1)
3 + 0.586988 (-1 + x)
3 + 0.123724 (-1 + x)
```

The line whose slope gives the central difference approximation passes through $(.5, f(.5))$ and $(1.5, f(1.5))$. Let's include it in the plot as well.

```
sec3[x_] = f[.5] + mcntrl[1, .5] (x - .5)
2.47487 + 0.586988 (-0.5 + x)
```

Finally we plot the curve along with all three secant lines. Note that the secant line whose slope came from the central difference approximation does a far better job of approximating the tangent line at (1, 3).

```
Plot[{f[x], secfwd[x], seccntrl[x], sec3[x]},
 {x, 0, 2}, PlotStyle → {Black, Red, Blue, Purple}]
```

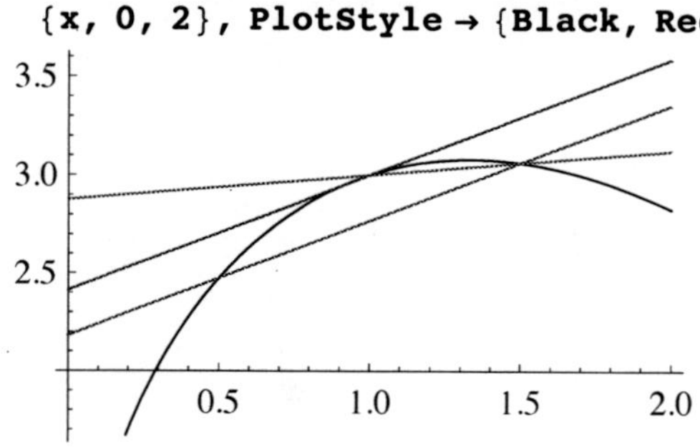

■ The Derivative

The preceding development motivates the definition of a function f', the *derivative* of f. The definition is

$$f'(x) = \lim_{h \to 0} \frac{f(x+h) - f(x)}{h}$$

for all x at which the defining limit exists. Note that values of f' give tangent line slopes on the graph of $y = f(x)$; that is,

$$f'(a) = m_{\tan} \text{ at } (a, f(a)).$$

A good approximation of $f'(x)$ can usually be computed by means of a central difference approximation. We'll refer to the resulting function as a numerical derivative:

$$\delta_h(x) = \frac{f(x+h) - f(x-h)}{2h}.$$

With this definition, we have

$$f'(x) \approx \delta_h(x) \text{ for small } h.$$

▪ Example 2.2.3

Consider the cubic polynomial

```
f[x_] := x³ - x
```

A numerical derivative of this function is

```
ndf[x_] = (f[x + .01] - f[x - .01]) / .02

50. (-0.02 - (-0.01 + x)³ + (0.01 + x)³)
```

Thus the slope of the tangent line to the graph of f at $(1, 0)$, for example, is approximately

```
ndf[1]

2.0001
```

and so an approximate tangent line there is $y = 2.0001\,(x - 1)$.

```
Plot[{f[x], ndf[1] (x - 1)}, {x, -1, 2}]
```

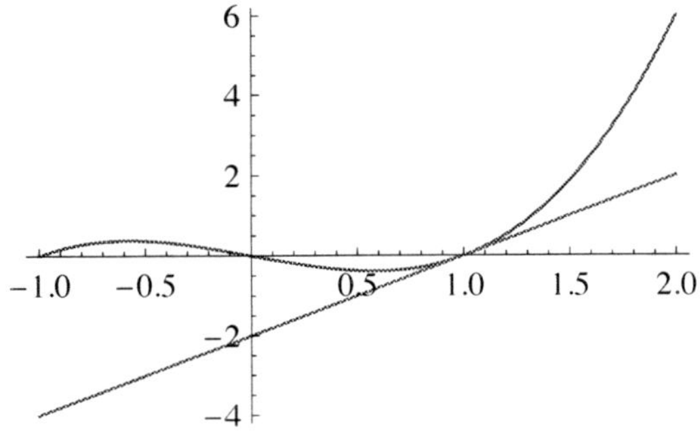

It is also instructive to look at the graphs of f and f' simultaneously. Keep in mind that the value of the quadratic f' is the slope of the graph of the cubic f. In particular, notice that $f'(x) > 0$ when $f(x)$ is increasing, and $f'(x) < 0$ when $f(x)$ is decreasing.

```
Plot[{f[x], ndf[x]}, {x, -1.5, 1.5}]
```

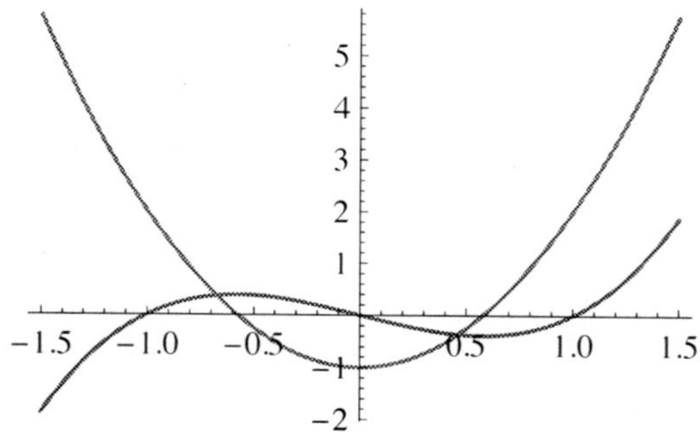

■ Exploring the Derivative Graph

The following "program" creates a `Manipulate` object in which each frame shows the graph of f on the interval $[x_{min}, x_{max}]$, the tangent line to the graph at a point $(a, f(a))$, and the graph of f' on $[x_{min}, a]$. As a takes on increasing values between x_{min} and x_{max}, we see the slope of the tangent line traced out as the graph of f'.

The tangent line slopes will be approximated by the numerical derivative

$$ndf(x) = \frac{f(x + .001) - f(x - .001)}{.002},$$

which we'll define first.

```
nd[fn_, x_] := 500 ((fn /. x → x + 0.001) - (fn /. x → x - 0.001));
slopeExplorer[fn_, {x_, xmin_, xmax_}, {y_, ymin_, ymax_}] :=
 Module[{f, ndf}, f[t_] = fn /. x → t; ndf[t_] = nd[fn, x] /. x → t;
  Manipulate[ Show[
    Plot[{f[x], f[a] + ndf[a] (x - a)}, {x, xmin, xmax}],
    Plot[ndf[x], {x, xmin - 0.01, a}, PlotStyle → {Thick}],
    Graphics[{Line[{{a, f[a]}, {a, ndf[a]}}], PointSize[Large],
      Red, Point[{a, f[a]}], Blue, Point[{a, ndf[a]}]}],
    PlotRange → {ymin, ymax}], {a, xmin, xmax}]]
```

The following is an example with the graph of $y = x^3 - x$.

```
slopeExplorer[x^3 - x, {x, -1.5, 1.5}, {y, -2, 3}]
```

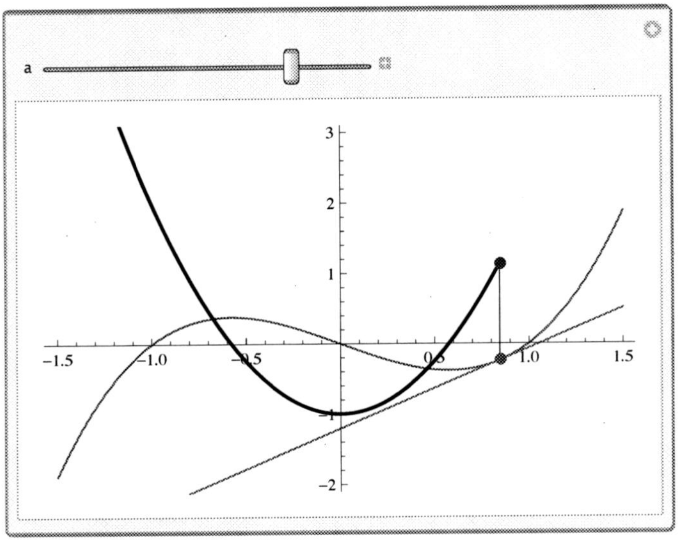

■ Symbolic Computation of Exact Derivatives

Mathematica has the extraordinary ability to compute symbolically the derivative of essentially any function we care to consider. There are various ways to ask *Mathematica* to compute a derivative. For a defined function, such as

```
g[x_] := x Sin[3 x]
```

we can appeal directly to the definition of the derivative:

```
Limit[(g[x + h] - g[x])/h, h → 0]
3 x Cos[3 x] + Sin[3 x]
```

We can also compute the derivative by entering either of these:

```
g′[x]
3 x Cos[3 x] + Sin[3 x]
D[g[x], x]
3 x Cos[3 x] + Sin[3 x]
```

The second of these is just the "`InputForm`" of the "`StandardForm`" command

```
∂x g[x]
3 x Cos[3 x] + Sin[3 x]
```

(The symbol "∂" is typed as ESC pd ESC, or you can click its button on the BasicInput palette.)

This third method is particularly useful for differentiating *expressions*, as opposed to functions. For example,

```
∂x (x^3 Sqrt[x^2 + 1])
```

$$\frac{x^4}{\sqrt{1+x^2}} + 3\,x^2\,\sqrt{1+x^2}$$

We can also compute derivatives of expressions that contain symbolic parameters.

```
∂t (x t^3 - ⅇ^(k t))
```

$$-e^{k\,t}\,k + 3\,t^2\,x$$

```
∂x (x^p)
```

$$p\,x^{-1+p}$$

When *Mathematica* computes a derivative symbolically, it is really just applying the very same differentiation rules that you apply when computing derivatives by hand. Notice that *Mathematica* "knows" the basic differentiation rules:

```
∂x {c u[x], u[x] + v[x], u[x] v[x], u[x]/v[x], u[x] v[x]} // TableForm
```

```
c u′[x]
u′[x] + v′[x]
v[x] u′[x] + u[x] v′[x]
u′[x]/v[x] - u[x] v′[x]/v[x]^2
v[x] u′[x] + u[x] v′[x]
```

By repeatedly applying these and other rules, *Mathematica* is able to compute very complicated derivatives like this, for example:

```
∂x ((x Cos[x] + Sqrt[x^2 + 1])/(Sin[x^2] Cos[x])) // Simplify
```

$$\frac{1}{\sqrt{1+x^2}}\,\mathrm{Csc}[x^2]\,\left(\sqrt{1+x^2} - 2\,x\,\left(1+x^2+x\,\sqrt{1+x^2}\,\mathrm{Cos}[x]\right)\,\mathrm{Cot}[x^2]\,\mathrm{Sec}[x] + \mathrm{Sec}[x]\,\left(x+\mathrm{Tan}[x]+x^2\,\mathrm{Tan}[x]\right)\right)$$

◆ Exercises

1. Plot the function $f(x) = x^2 - 1$ for $-2 \le x \le 2$ together with its tangent lines at $x = -1, 0,$ and 1. Use a central difference approximation with $h = .01$ for the tangent line slopes.

2. Plot the function $f(x) = \sin x$ for $-1 \le x \le 7$ together with its tangent lines at $x = 0, \pi/2, \pi,$ $3\pi/2$, and 2π. Use a central difference approximation with $h = .01$ for the tangent line slopes.

In Exercises 3-5, for the given function f and number a, compare the values of

a) the (forward) difference quotient $\dfrac{f(a+0.1) - f(a)}{0.1}$,

b) the central difference quotient $\dfrac{f(a+0.1) - f(a-0.1)}{0.2}$,

c) the exact derivative value $f'(a)$.

3. $f(x) = 3x^2 - x + 2,\ a = 1$
4. $f(x) = e^x,\ a = 1$
5. $f(x) = \sin x,\ a = \pi/3$
6. For each of the following functions, use `slopeExplorer` to explore the relationship between the graphs of f and f' on the indicated interval. Then, from what you observe in the graph *and without computing* $f'(x)$, hazard a guess as to exactly what $f'(x)$ is.

   a) $f(x) = \sin x,\ 0 \le x \le 2\pi$ b) $f(x) = \cos x,\ 0 \le x \le 2\pi$

   c) $f(x) = \sin 2x,\ 0 \le x \le 2\pi$ d) $f(x) = e^x,\ -2 \le x \le 2$ e) $f(x) = e^{-x},\ -2 \le x \le 2$
7. For each function in Exercise 6, compute its difference quotient $m_x(h) = (f(x+h) - f(x))/h$. Then compute $\lim_{h \to 0} m_x(h)$ to obtain the derivative of f.
8. Find all x in the interval $[0, 2]$ where the tangent line to the graph of $f(x) = x^3 - x^2 - x$ is parallel to the secant line through $(0, 0)$ and $(2, 2)$. Plot the graph of f, the secant line, and the parallel tangent line(s) on the interval $[0, 2]$. Can you think of an example of a function f and an interval $[a, b]$ for which no such x could be found?
9. Below are four sets of four functions each. Use ∂_x to compute the derivative of each function. After doing all three sets, try to observe a pattern, and state from it a general rule about the derivative of the composition of two functions.

   a) $x^3,\ \sin x,\ \sin x^3,\ \sin^3 x$ b) $\tan x,\ e^x,\ e^{\tan x},\ \tan e^x$

   c) $1/x,\ \cos x,\ \cos(1/x),\ 1/\cos x$ d) $\ln x,\ 1/x,\ 1/\ln x,\ \ln(1/x)$
10. **String graphs.** Interesting pictures arise from plotting a large number of tangent lines to a given curve. For example,

```
f[x_] := x^2;
Plot[Evaluate[Table[f[a] + f'[a] (x - a), {a, -2, 2, .1}]],
 {x, -2, 2}, PlotRange → {-4, 4}]
```

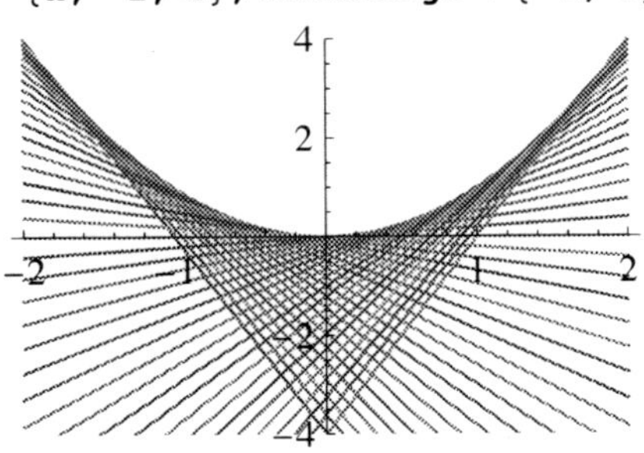

Experiment with other functions, such as x^3, x^4, and $\cos x$. You may want to change the interval, the `PlotRange` specification, or the number of lines plotted.

2.3 Higher-order Derivatives

It is also easy to compute higher-order derivatives with *Mathematica*. For example, consider

```
g[x_] := x^2 Cos[a x]
```

To compute the second derivative of g, we can enter

```
g''[x]
2 Cos[a x] - a^2 x^2 Cos[a x] - 4 a x Sin[a x]
```

or

```
D[g[x], {x, 2}]
2 Cos[a x] - a^2 x^2 Cos[a x] - 4 a x Sin[a x]
```

or

```
∂x,x g[x]
2 Cos[a x] - a^2 x^2 Cos[a x] - 4 a x Sin[a x]
```

Third- and higher-order derivatives are computed analogously. For example, the third through fifth derivatives are

```
g'''[x]
D[g[x], {x, 4}]
∂x,x,x,x,x g[x]

-6 a^2 x Cos[a x] - 6 a Sin[a x] + a^3 x^2 Sin[a x]
-12 a^2 Cos[a x] + a^4 x^2 Cos[a x] + 8 a^3 x Sin[a x]
10 a^4 x Cos[a x] + 20 a^3 Sin[a x] - a^5 x^2 Sin[a x]
```

We can even compute a list of derivatives as follows:

```
Table[D[g[x], {x, k}], {k, 0, 8}] // TableForm

x^2 Cos[a x]
2 x Cos[a x] - a x^2 Sin[a x]
2 Cos[a x] - a^2 x^2 Cos[a x] - 4 a x Sin[a x]
-6 a^2 x Cos[a x] - 6 a Sin[a x] + a^3 x^2 Sin[a x]
-12 a^2 Cos[a x] + a^4 x^2 Cos[a x] + 8 a^3 x Sin[a x]
10 a^4 x Cos[a x] + 20 a^3 Sin[a x] - a^5 x^2 Sin[a x]
30 a^4 Cos[a x] - a^6 x^2 Cos[a x] - 12 a^5 x Sin[a x]
-14 a^6 x Cos[a x] - 42 a^5 Sin[a x] + a^7 x^2 Sin[a x]
-56 a^6 Cos[a x] + a^8 x^2 Cos[a x] + 16 a^7 x Sin[a x]
```

The Second Derivative and Concavity

The first and second derivatives of a function f have simple interpretations in terms of the graph of f. The sign of $f'(x)$ determines where the function f is increasing or decreasing. The sign of $f''(x)$ determines where the graph of f is concave up or concave down. This is illustrated as follows with the function $f(x) = x^4 - 2x^2$. First we'll graph f together with f'.

```
f[x_] := x^4 - 2 x^2
Plot[{f[x], f'[x]}, {x, -2, 2}, PlotRange → {-5, 7}]
```

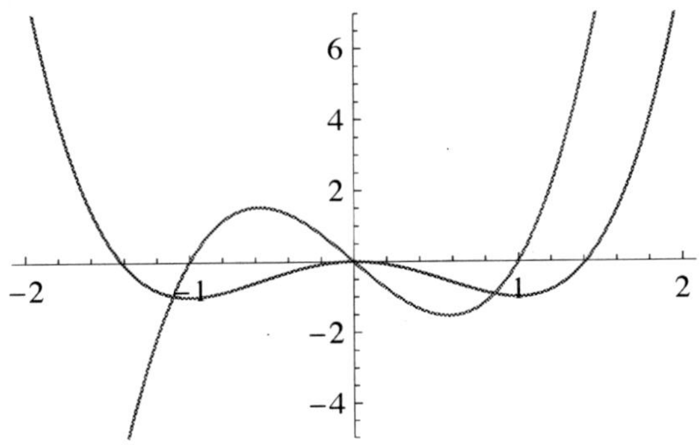

Notice that $f'(x) > 0$ wherever $f(x)$ is increasing, $f'(x) < 0$ wherever $f(x)$ is decreasing, and $f'(x) = 0$ where each local maximum or minimum value of f occurs. These facts are not at all surprising in light of the fact that f' gives the slope of the graph of f. Now let's look at the graph of f together with the graph of f''.

```
Plot[{f[x], f''[x]}, {x, -2, 2}, PlotRange → {-5, 7}]
```

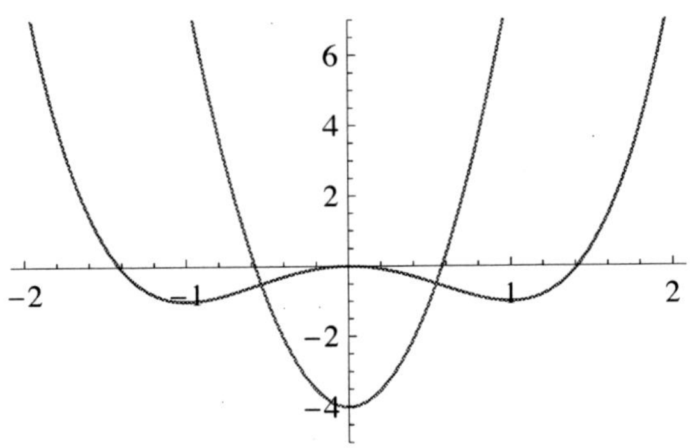

Notice that $f''(x) > 0$ wherever the graph of f is concave up, $f''(x) < 0$ wherever the graph of f is concave down. The reason for this is that the concavity of the graph of f is determined by whether the slope $f'(x)$ is increasing or decreasing.

You should spend a considerable amount of time contemplating pictures such as these.

■ Exploring the Graph of the Second Derivative

The following program is a simple modification of the `slopeExplorer` program in the previous section. This program traces out the graph of the second derivative as the tangent line moves along the graph of the function.

```
concavityExplorer[fn_, {x_, xmin_, xmax_}, {y_, ymin_, ymax_}] :=
 Module[{f}, f[t_] = fn /. x → t;
  Manipulate[Show[
    Plot[{f[x], (f[a]) + f′[a] (x - a)},
     {x, xmin, xmax}, PlotRange → All],
    Plot[f″[x], {x, xmin - .01, a}, PlotStyle → Thick],
    Graphics[{Gray, Line[{{a, f[a]}, {a, f″[a]}}],
      PointSize[Large], Red, Point[{a, f[a]}],
      Blue, Point[{a, f″[a]}]}], PlotRange → {ymin, ymax}],
   {a, xmin, xmax}]]
```

The important thing to observe as you watch the animation is that the second derivative is positive when the tangent line turns in a counter-clockwise direction (indicating that the slope of the graph is increasing), and the second derivative is negative when the tangent line turns in a clockwise direction (indicating that the slope of the graph is decreasing).

The following provides a nice example.

```
concavityExplorer[x^2 Sin[x / 2] / 10, {x, -2 π, 2 π}, {y, -2, 2}]
```

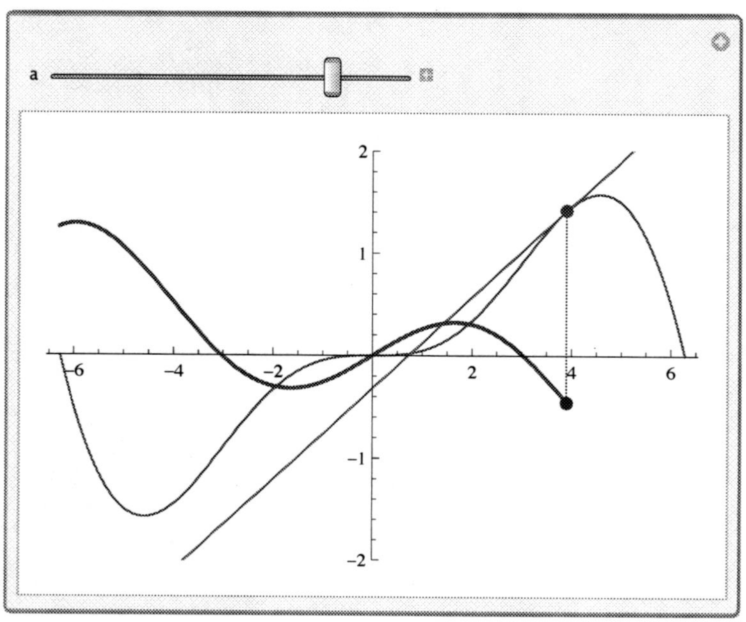

◆ Exercises

For the functions in Exercises 1-4, first use `slopeExplorer` to plot the function and its derivative on the suggested interval, noting the correspondence between the sign of the derivative and the increasing/decreasing nature of $f(x)$. Then use `concavityExplorer` to plot the function and its second derivative, noting the correspondence between the sign of the second derivative and the concavity of the graph of f.

1. $f(x) = \dfrac{1}{x^2 + 1}$ on $[-5, 5]$

2. $f(x) = \dfrac{x}{x^2 + 1}$ on $[-10, 10]$

3. $f(x) = \tan^{-1} x$ on $[-10, 10]$

4. $f(x) = \sqrt{x} - x$ on $[0, 3]$

For the functions in 5 and 6, graph the function along with its second derivative. Then solve for each x where $f''(x)$ changes sign. The points on the graph where $f''(x)$ changes sign are **inflection points**. (For the function in #6, enter $x^{5/3}$ as `Sign[x] Abs[x]^5/3`.)

5. $f(x) = x^5 - 2x^3$

6. $f(x) = x^{5/3}(2 - x)^2$

For the functions in 7-10, compute a `Table` containing $f^{(n)}(0)$ for $n = 1, \ldots, 5$, where $f^{(n)}$ denotes the nth derivative of f. By observing a pattern in these numbers, express $f^{(n)}(0)$ in terms of n.

7. $f(x) = x e^{-x}$

8. $f(x) = x \cos x$

9. $f(x) = \dfrac{1}{1 + x}$

10. $f(x) = \sqrt{x + 1}$

3 Applications of the Derivative

We have seen that the derivative of f is a function whose value at $x = a$ gives the slope of the graph of f at the point $(a, f(a))$. This geometric idea has numerous variations and lends itself to many applications. In this chapter we will explore some of the many applications of the derivative with the help of *Mathematica*.

3.1 Velocity, Acceleration, and Rectilinear Motion

Let t be a variable representing the time elapsed since some reference time $t = 0$, and imagine a particle moving along a straight-line path in some way. Our interest here is in the function $s(t)$ that gives the position at time t of the moving particle.

■ Average and Instantaneous Velocity

Average velocity over a time interval $a \le t \le b$ is defined to be the change in position divided by the change in time:

$$v_{\text{av}} = \frac{s(b) - s(a)}{b - a}.$$

Notice that v_{av} is simply the slope of the secant line through $(a, s(a))$ and $(b, s(b))$.

▪ Example 3.1.1a

Consider a particle moving along a straight line with position

```
s[t_] := (t - 2)^3 - 2 t + 8
```

for $0 \le t \le 4$. Let's plot the graph of s along with secant lines for the time intervals $0 \le t \le 3$ and $0 \le t \le 4$. The slopes of these secant lines are the average velocities of the particle over the respective time intervals.

```
Show[Plot[s[t], {t, 0, 4}, PlotStyle → Black],
 Plot[s[0] + (s[3]-s[0])/(3 - 0) (t - 0), {t, 0, 3}, PlotStyle → Blue],
 Plot[s[0] + (s[4]-s[0])/(4 - 0) (t - 0), {t, 0, 4}, PlotStyle → Red]]
```

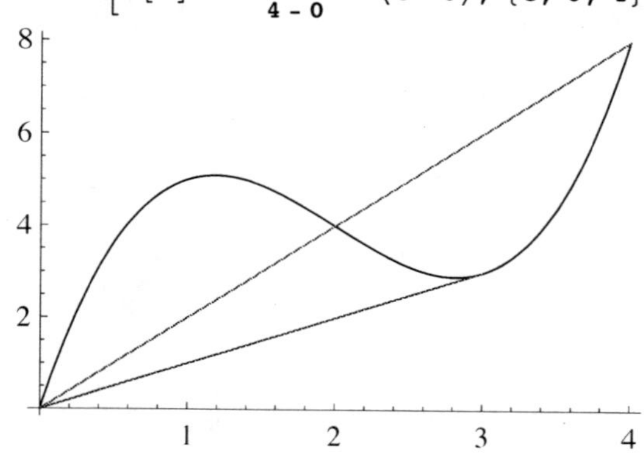

(Instantaneous) velocity at a time t is defined to be the limit of the average velocities over intervals $[t, t+h]$ as $h \to 0$. Thus velocity is the derivative of position:

$$v(t) = s'(t) = \lim_{h \to 0} \frac{s(t+h) - s(t)}{h}.$$

We also say that velocity is the (instantaneous) *rate of change* in position.

Example 3.1.1b

Let's plot the position function $s(t) = (t-2)^3 - 2t + 8$ (from Example 3.1.1a) along with the velocity $v(t) = s'(t)$.

```
Plot[{s[t], s'[t]}, {t, 0, 4}]
```

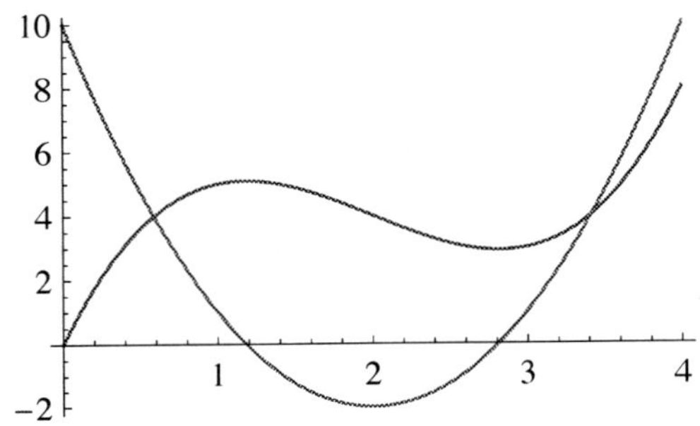

Notice that in this example, $s(t)$ is always positive, reflecting the fact that the position of the particle is strictly increasing; i.e., the particle is always moving forward. Also, for $0 \le t \le 2$, the velocity is decreasing, and for for $2 \le t \le 4$, the velocity is increasing.

Acceleration

Just as velocity is the rate of change in position, acceleration is the rate of change in velocity. Thus acceleration is the second derivative of position:

$$a(t) = v'(t) = s''(t).$$

Example 3.1.1c

Again consider the position function $s(t) = (t-2)^3 - 2t + 8$ for $0 \le t \le 4$. Let's add the graph of the acceleration to the plot of Example 3.1.1b.

```
Plot[{s[t], s'[t], s''[t]}, {t, 0, 4}]
```

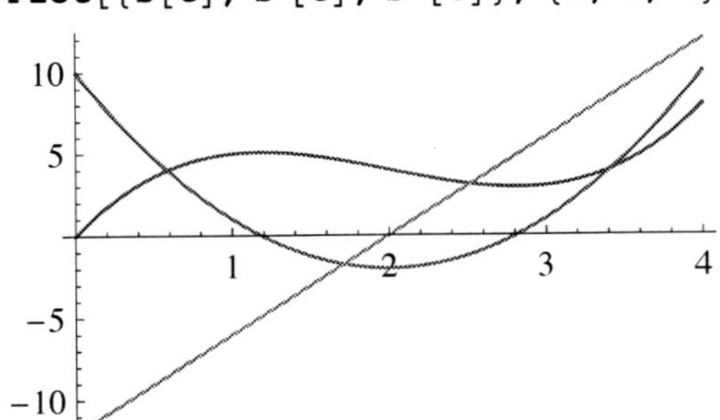

Notice that velocity is decreasing wherever acceleration is negative, and velocity is increasing wherever acceleration is positive.

■ Simulating Motion

Mathematica's ability to create animations lets us simulate motion and simultaneously plot position, velocity, and/or acceleration graphs. The following program, named `motion`, simulates the rectilinear motion of a red ball and simultaneously plots the position function and a portion of its tangent line.

The first argument provided to `motion` will be the position function in the form of an expression. The second argument is a four-element list that specifies the variable, its starting and ending values, and the time-step between frames. The third argument is a two-element list that specifies the positions at the bottom and top of the plot.

```
motion[positionf_, {t_, tmin_, tmax_}, {htmin_, htmax_}] :=
 Module[{track, ball, level, pt, grf, s, c, d},
  s[tt_] := positionf /. t → tt;
  c = tmin - (tmax - tmin) / 4; d = tmin - (tmax - tmin) / 5;
  track = Graphics[{Line[{{d, htmin}, {d, htmax}}]}];
  ball[y_] = Graphics[{PointSize[.033], Red, Point[{d, y}]}];
  level[tt_, y_] = Graphics[{Line[{{d, y}, {tt, y}}]}];
  pt[tt_, y_] = Graphics[{PointSize[.02], Blue, Point[{tt, y}]}];
  grf = Plot[s[t], {t, 0, tmax}, PlotStyle → Darker[Gray]];
  Animate[ Show[grf, track,
    Plot[s[t], {t, tmin - .001, t2}, PlotStyle → Thick],
    Plot[Evaluate[s[t2] + s'[t2] (t - t2)], {t, t2 - d, t2 + d},
     PlotStyle → Darker[Green]], level[t2, s[t2]], ball[s[t2]],
    pt[t2, s[t2]], Ticks → {Range[Floor[tmin], tmax], Automatic},
    PlotRange → {{c, tmax}, {htmin, htmax}}], {t2, tmin, tmax}]]
```

To illustrate `motion`, let's use the same position function as in Examples 3.1.abc.

```
s[t_] := (t - 2)^3 - 2 t + 8
```

To track the motion of the ball and plot its height over the interval $0 \le t \le 4$, enter

```
motion[s[t], {t, 0, 4.1}, {-.5, 8.5}]
```

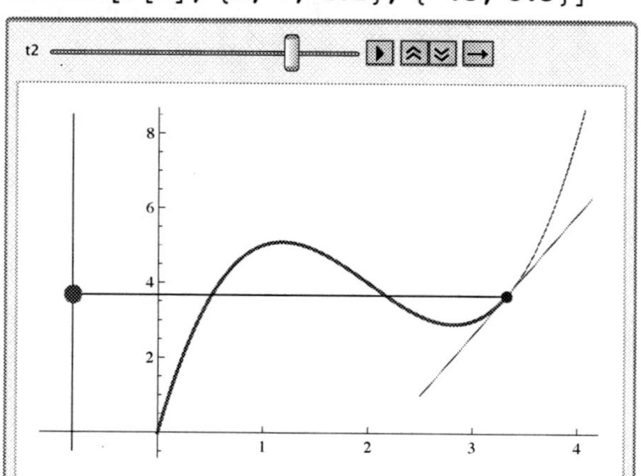

In the figure below, many of the resulting frames are superimposed.

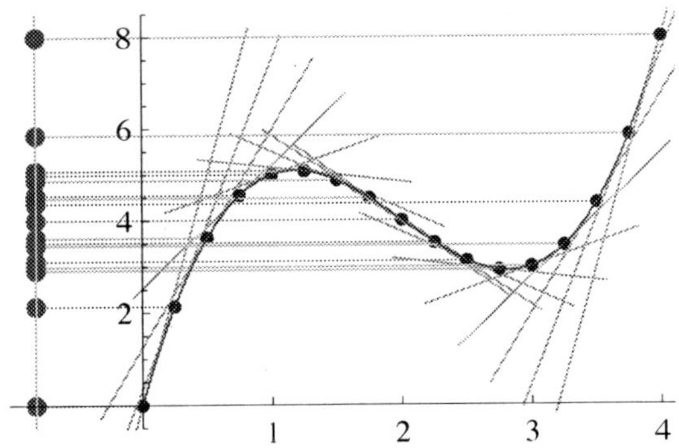

Another nice example is *simple harmonic motion*:

```
motion[Sin[t], {t, 0, 4 π}, {-1.3, 1.3}]
```

◆ Exercises

For each position function, plot the position, velocity, and acceleration on the indicated interval. Then simulate the motion using the `motion` program above.

1. $s(t) = \dfrac{t-3}{(t-3)^4+1}, \quad 0 \le t \le 6$

2. $s(t) = \sin(\pi t)\cos(5\pi t), \quad 0 \le t \le 2$

3. $s(t) = e^{-t/2}\sin \pi t, \quad 0 \le t \le 6$

Provided we ignore air resistance, the height (in feet) of a free-falling object, under the influence of gravity, is described approximately by

$$h(t) = -16\,t^2 + v_0\,t + h_0 \,,$$

where v_0 is the velocity at time $t = 0$ (or *initial velocity*), and h_0 is the height at time $t = 0$ (or *initial height*). For each combination of initial velocity and initial height in 4-6,

a) plot the height and velocity for $0 \le t \le T$, where T is the time when the height becomes 0;

b) find the maximum height that the object attains;

c) use the `motion` program to simulate the motion.

4. $v_0 = 100\,\text{ft/sec}, \quad h_0 = 0$

5. $v_0 = 64\,\text{ft/sec}, \quad h_0 = 25$

6. $v_0 = 0\,\text{ft/sec}, \quad h_0 = 100$

3.2 Implicit Differentiation and Related Rates

When two or more variables are related through the requirement that a given equation in them is always true, it is natural to think of any one of the variables as being dependent upon the others. Thus we can inquire about the rate of change in one variable with respect to another, even though the relation typically is not one given by a function, per se. The key to such questions is always one form or another of the chain rule.

■ Implicit Differentiation

Suppose that we are interested in studying the graph of an equation such as

```
eqn[x_, y_] := ((x^2 + 2 y^2)^2 == x - 3 x y + y)
```

which defines neither variable as a function of the other. *Mathematica*'s `ContourPlot` function can be used to plot the graph of such an equation as follows:

```
curve =
 ContourPlot[Evaluate[eqn[x, y]], {x, -1, 1.5}, {y, -1, 1}]
```

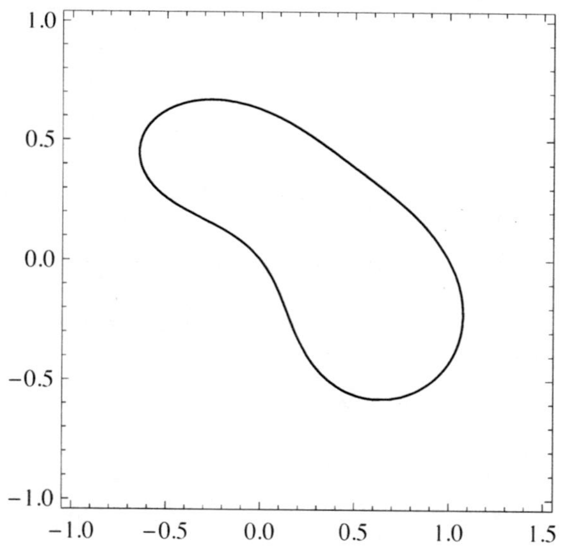

In cases such as this, in which it is either difficult or impossible to solve for either variable explicitly, finding the derivatives $\frac{dy}{dx}$ or $\frac{dx}{dy}$ is typically *not* difficult. We can find each of these derivatives, directly from the equation, by **implicit differentiation**. The process for finding $\frac{dy}{dx}$ is as follows.

1) Find the derivative of each side of the equation "implicitly" with respect to x.

2) Solve the resulting equation for $\frac{dy}{dx}$ in terms of x and y.

■ Example 3.2.1

Find each of $\frac{dy}{dx}$ and $\frac{dx}{dy}$ for the equation whose graph is plotted above. Then find each of the points on the curve where the tangent line is either horizontal or vertical.

We differentiate the equation itself with respect to x as follows.

```
yprimeEq = D[eqn[x, y[x]], x]
```

$$2\left(x^2+2\,y[x]^2\right)\,(2\,x+4\,y[x]\,y'[x]) == 1-3\,y[x]+y'[x]-3\,x\,y'[x]$$

Note the substitution of `y[x]` for `y` so that *Mathematica* doesn't treat `y` as a constant. The second step is to solve for $y'(x)$:

```
sol = Solve[yprimeEq, y'[x]] // Simplify
```

$$\left\{\left\{y'[x]\to\frac{1-4\,x^3-3\,y[x]-8\,x\,y[x]^2}{-1+3\,x+8\,x^2\,y[x]+16\,y[x]^3}\right\}\right\}$$

Now to convert the result to the form of an expression in x and y, we'll enter

```
dydx = y'[x] /. First[sol] /. y[x] → y
```

$$\frac{1-4\,x^3-3\,y-8\,x\,y^2}{-1+3\,x+8\,x^2\,y+16\,y^3}$$

To differentiate with respect to y, we enter

```
xprimeEq = D[eqn[x[y], y], y]
```

$$2\left(2\,y^2+x[y]^2\right)\,(4\,y+2\,x[y]\,x'[y]) == 1-3\,x[y]+x'[y]-3\,y\,x'[y]$$

Note the substitution of `x[y]` for `x` so that *Mathematica* doesn't treat `x` as a constant. Now we solve for $x'(y)$ in an analogous manner to the above.

```
sol = Solve[xprimeEq, x'[y]] // Simplify
```

$$\left\{\left\{x'[y]\to\frac{1-16\,y^3-3\,x[y]-8\,y\,x[y]^2}{-1+3\,y+8\,y^2\,x[y]+4\,x[y]^3}\right\}\right\}$$

```
dxdy = x'[y] /. First[sol] /. x[y] → x
```

$$\frac{1-3\,x-8\,x^2\,y-16\,y^3}{-1+4\,x^3+3\,y+8\,x\,y^2}$$

Now that we have each of the required derivatives, we can find the points on the curve where the tangent line is either horizontal or vertical. Points where the tangent line is horizontal will be the real solutions among these:

```
horzpts = NSolve[{eqn[x, y], dydx == 0}, {x, y}]

{{x → 0.63572, y → -0.580506}, {x → -0.261366, y → 0.670179},
 {x → 0.485392 - 0.511037 i, y → 0.363355 + 0.216327 i},
 {x → 0.485392 + 0.511037 i, y → 0.363355 - 0.216327 i},
 {x → -0.287186 - 0.393079 i, y → 0.169315 + 0.0938541 i},
 {x → -0.287186 + 0.393079 i, y → 0.169315 - 0.0938541 i},
 {x → -0.0520502 - 0.502452 i, y → -0.244173 - 0.708673 i},
 {x → -0.0520502 + 0.502452 i, y → -0.244173 + 0.708673 i}}

pts = {x, y} /. horzpts[[{1, 2}]]

{{0.63572, -0.580506}, {-0.261366, 0.670179}}
```

Similarly, points where the tangent line is vertical will be the real solutions among these:

```
vertpts = NSolve[{eqn[x, y], dxdy == 0}, {x, y}]

{{x → 1.07391, y → -0.221866}, {x → -0.645952, y → 0.448291},
 {x → -0.381734 - 1.01893 i, y → -0.00184328 - 0.340933 i},
 {x → -0.381734 + 1.01893 i, y → -0.00184328 + 0.340933 i},
 {x → 0.367024 - 0.386544 i, y → 0.420007 + 0.310331 i},
 {x → 0.367024 + 0.386544 i, y → 0.420007 - 0.310331 i},
 {x → 0.134064 - 0.0844787 i, y → -0.198043 + 0.269498 i},
 {x → 0.134064 + 0.0844787 i, y → -0.198043 - 0.269498 i}}

pts = Join[pts, {x, y} /. vertpts[[{1, 2}]]]

{{0.63572, -0.580506}, {-0.261366, 0.670179},
 {1.07391, -0.221866}, {-0.645952, 0.448291}}
```

Finally, the following is a plot of the curve that shows the points we've found.

```
Show[curve, Graphics[{PointSize[.03], Blue, Map[Point, pts]}]]
```

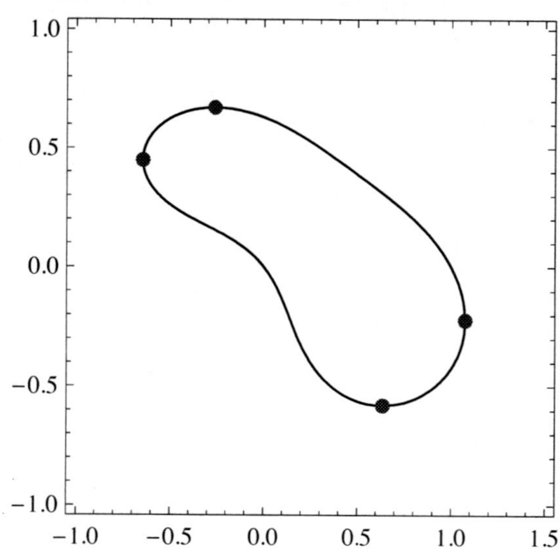

Related Rates Problems

When two or more quantities that are themselves functions of time are related through some equation, a relationship among the rates of change in those quantities can be obtained by implicit differentiation with respect to t.

Example 3.2.2

A particle is moving along the curve $x^3 + x - y^3 - y = 8$ *in such a way that the x-coordinate of the particle's position is changing at a constant rate* $x'(t) = 1$. *How fast is the y-coordinate of the particle's position changing when the particle is at* (2, 1) *?*

```
ContourPlot[x^3 + x - y^3 - y == 8, {x, -3, 4}, {y, -3, 4},
 Epilog → {PointSize[Large], Red, Point[{2, 1}]}]
```

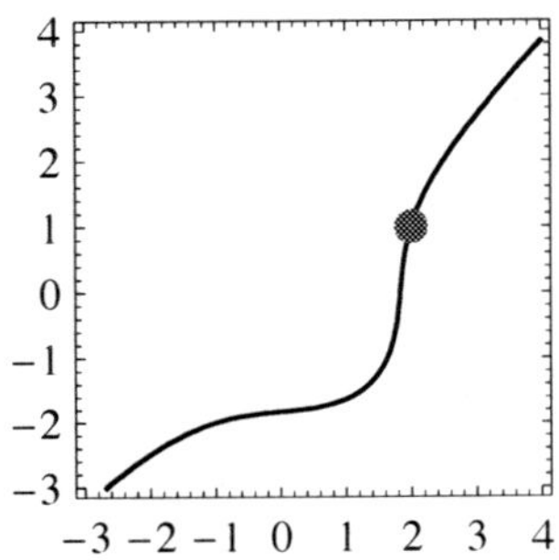

To solve this problem, we first enter the equation of the path, using $x(t)$ and $y(t)$ to indicate that x and y are to be treated as functions of t. Then we'll differentiate implicitly with respect to t.

```
eqn := x[t]^3 + x[t] - y[t]^3 - y[t] == 8;
dtEqn = D[eqn, t]
```

$$x'[t] + 3\,x[t]^2\,x'[t] - y'[t] - 3\,y[t]^2\,y'[t] == 0$$

Now we solve for $y'(t)$:

```
dydt = y'[t] /. (Solve[dtEqn, y'[t]] // Flatten)
```

$$\frac{x'[t] + 3\,x[t]^2\,x'[t]}{1 + 3\,y[t]^2}$$

and then substitute $x'(t) = 1$ into that expression:

```
dydt = dydt /. x'[t] → 1
```

$$\frac{1 + 3\,x[t]^2}{1 + 3\,y[t]^2}$$

Finally, the rate of change in y at the point (2, 1) is

```
dydt /. {x[t] → 2, y[t] → 1}
```

$$\frac{13}{4}$$

Formulas from geometry (as well as physics, chemistry, and elsewhere) often describe the relationship between two or more quantities that may be changing with time.

■ Example 3.2.3

One tenth of one cubic inch of oil is dropped gently onto the surface of a pan of water, quickly spreading out in all directions and taking on an approximately cylindrical shape. The radius of the oil is observed increasing at a rate of 1/2 in/s at the instant when the radius is 5 inches. Find the rate of change in the thickness of the oil at that instant.

The radius r and thickness y are related by the formula for the volume of a cylinder:

```
volEq := (π r[t]^2 y[t] == 1)
```

First we'll differentiate this equation with respect to t. Because of the volume formula, we will also substitute $1/(\pi r(t)^2)$ for $y(t)$.

```
yprimeEq = D[volEq, t] /. y[t] → 1/(π r[t]^2)
```

$$\frac{2\,r'[t]}{r[t]} + \pi\, r[t]^2\, y'[t] == 0$$

Next we'll solve for $y'(t)$, substituting $r(t) = 1/2$ into the result.

```
dydt = y'[t] /. (Solve[yprimeEq, y'[t]] // Flatten) /. r'[t] → 1/2
```

$$-\frac{1}{\pi\, r[t]^3}$$

Finally, we evaluate $y'(t)$ when $r(t) = 5$.

```
dydt /. {r[t] → 5}
% // N
```

$$-\frac{1}{125\,\pi}$$

-0.00254648

Thus the thickness is *decreasing* at $1/(125\,\pi) \approx .00255$ in/s when $r = 5$ inches.

◆ Exercises

1. Find and read the description of `ContourPlot` in the *Mathematica* Documentation Center.
2. For each of the following equations, plot the graph of the equation with `ContourPlot` and then use implicit differentiation to find the slope of the tangent line to the graph at the indicated point.

   a) $x\,y^2 - y\,x^3 = 2$ at $(0, 2)$ (Plot on $-5 \le x \le 5,\ -5 \le y \le 5$.)

   b) $2\sin(\pi y\cos(\pi x)) + 3\,x\,y = 2$ at $(1/3, 1)$ (Plot on $0 \le x \le 3,\ 0 \le y \le 2$.)

   c) $x\sin(\pi y) - y\cos(\pi x) = 1$ at $(2/3, 2)$ (Plot on $0 \le x \le \pi,\ 0 \le y \le \pi$.)

3. A particle moves along the parabola $y = x^2$ in such a way that the x-coordinate of the particle's position changes at a constant rate $x'(t) = 2$. Find the rate of change in the y-coordinate of the particle's position at the instant when the position of the particle is:

   a) $(-1, 1)$ b) $(0, 0)$ c) $(1, 1)$ d) $(2, 4)$

4. A particle moves along the top half of the circle $x^2 + y^2 = 1$ in such a way that the x-coordinate of the particle's position changes at a constant rate $x'(t) = 1$. Find the rate of change in the y-coordinate of the particle's position at the instant when the position of the particle is:

   a) $\left(-\sqrt{1/2}, \sqrt{1/2}\right)$ b) $(0, 1)$ c) $\left(\sqrt{3}/2, 1/2\right)$

5. For each of the following equations, differentiate implicitly and then solve for the indicated rate. The symbol "k" is always a constant.

   a) $V(t) = \frac{4\pi}{3} r(t)^3, \ \frac{dr}{dt}$ b) $V(t) = \frac{\pi}{3} r(t)^2 h(t), \ \frac{dh}{dt}$ c) $p(t)\,V(t) = k\,T(t), \ \frac{dp}{dt}$

   d) $A(t) = 2\pi r(t)\,(r(t) + h(t)), \ \frac{dr}{dt}$ e) $\frac{\sin\theta(t)}{\sin\phi(t)} = k, \ \frac{d\theta}{dt}$

3.3 Linear and Quadratic Approximation

■ Linearization

Consider the following problem:

> *Given a differentiable function f and a number a in its domain, find the linear function λ_a that best approximates f near a in the sense that*
>
> $$\lambda_a(a) = f(a) \quad \text{and} \quad \lambda_a'(a) = f'(a).$$

In other words, we are looking for the linear function whose graph passes through $(a, f(a))$ with the same slope as the tangent line to the graph of f there. Since λ_a is linear, we can write $\lambda_a(x) = m\,x + b$, and so our requirements on λ_a become

$$m\,a + b = f(a) \quad \text{and} \quad m = f'(a)$$

Now we solve for m and a to get $m = f'(a)$ and $b = f(a) - a\,f'(a)$. This gives us, after a little rearranging, the *linearization* of f at $x = a$:

$$\lambda_a(x) = f(a) + f'(a)\,(x - a).$$

Of course, this is nothing more than the function whose graph is the tangent line to the graph of f at $(a, f(a))$.

■ Example 3.3.1

Let's find and plot the linearization of $f(x) = x^3$ at $x = 3/4$.

```
f[x_] := x^3
λ[x_] = f[3 / 4] + f′[3 / 4] (x - 3 / 4)
```

$$\frac{27}{64} + \frac{27}{16}\left(-\frac{3}{4} + x\right)$$

```
Plot[{f[x], λ[x]}, {x, 0, 1},
 PlotRange → {-.1, 1}, PlotStyle → {{}, Red}]
```

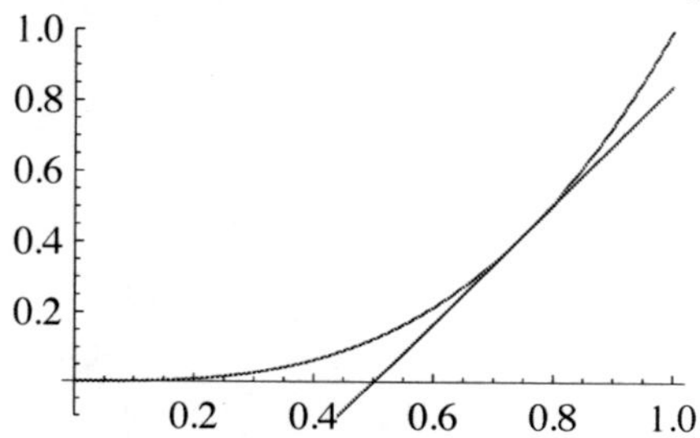

To get an idea of the range of values of x for which $\lambda_{3/4}(x)$ gives a good approximation to $f(x)$, let's create a plot that shows both graphs on the interval [.6, .9] and a graph of the error $|f(x) - \lambda_{3/4}(x)|$ on the same interval.

```
GraphicsRow[
 {Plot[{f[x], λ[x]}, {x, .6, .9}, PlotStyle → {{}, Red}],
  Plot[Abs[f[x] - λ[x]], {x, .6, .9}]}]
```

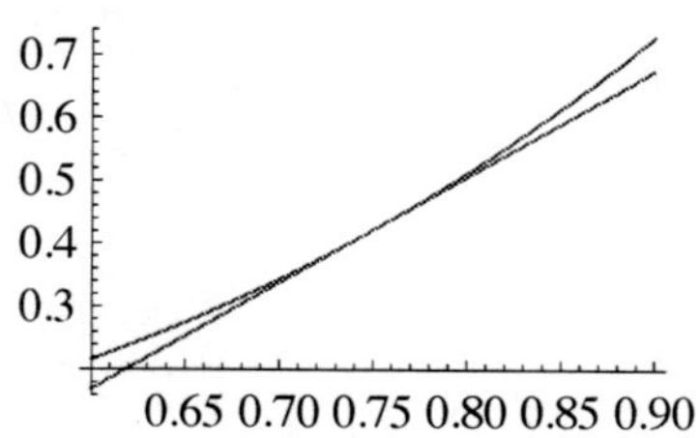

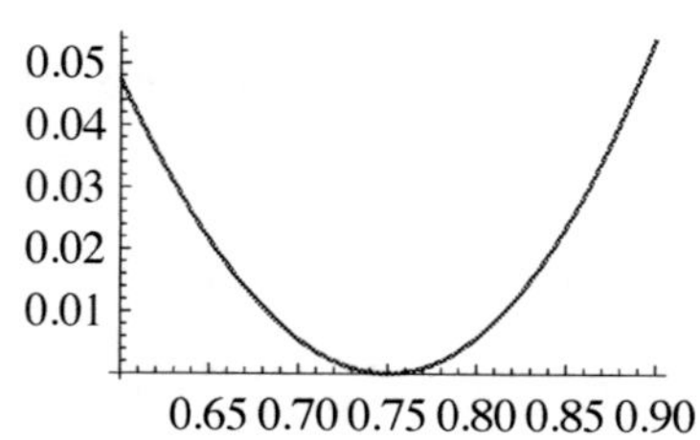

So it appears that $|f(x) - \lambda_{3/4}(x)| < 0.05$ whenever $|x - 3/4| \le 0.15$. Finally, let's make a similar plot for the interval [.7, .8].

```
GraphicsRow[
 {Plot[{f[x], λ[x]}, {x, .7, .8}, PlotStyle → {{}, Red}],
  Plot[Abs[f[x] - λ[x]], {x, .7, .8}]}]
```

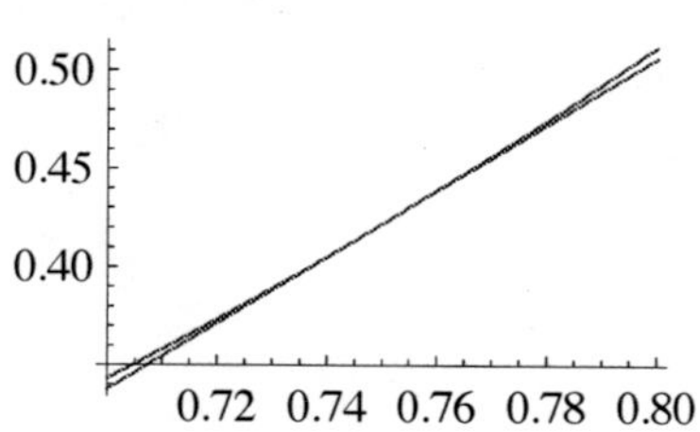

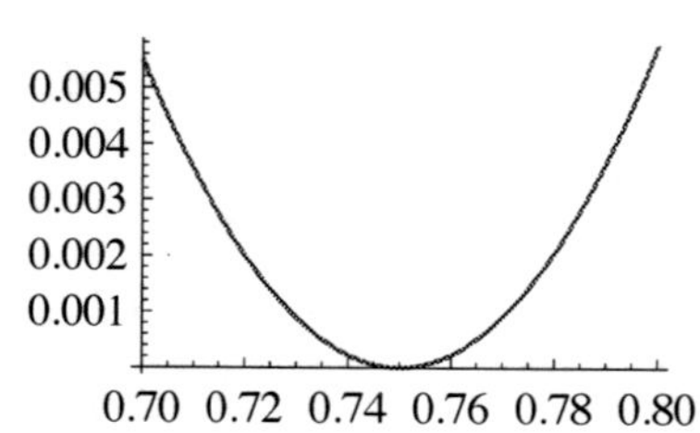

So it appears that $|f(x) - \lambda_{3/4}(x)| < 0.0055$ whenever $|x - 3/4| \le 0.05$.

▪ Example 3.3.2

Use linear approximation to approximate cos(1).

Since 1 is fairly close to $\pi/3$, we will linearize the cosine function about $\pi/3$:

```
λ[x_] = Cos[π / 3] - Sin[π / 3] (x - π / 3)
```

$$\frac{1}{2} - \frac{1}{2}\sqrt{3}\left(-\frac{\pi}{3} + x\right)$$

The resulting approximation of cos(1) is

```
λ[1.]
0.540874
```

which agrees to two decimal places with

```
Cos[1.]
0.540302
```

▪ Quadratic Approximation

Our approach here will parallel our approach to the linear approximation. Consider the following problem.

Given a twice-differentiable function f and a number a in its domain, find the quadratic function q_a that best approximates f near a in the sense that

$$q_a(a) = f(a),\ \ q_a'(a) = f'(a) \ \text{ and } \ q_a''(a) = f''(a)\ .$$

To make the derivation proceed more smoothly, we'll look for $q_a(x)$ in the form

$$q_a(x) = c_0 + c_1(x-a) + c_2(x-a)^2.$$

Because of this, our requirements on q_a become

$$c_0 = f(a),\ c_1 = f'(a),\ \text{and}\ 2\,c_2 = f''(a).$$

So we arrive at

$$q_a(x) = f(a) + f'(a)\,(x-a) + \frac{1}{2}\,f''(a)\,(x-a)^2.$$

Notice that the first two terms are precisely the linear approximation $\lambda_a(x)$ from before.

▪ Example 3.3.3

Let's find and plot the quadratic approximation of $f(x) = \cos x$ at $a = 0$.

```
f[x_] := Cos[x]
a := 0
q[x_] = f[a] + f′[a] (x - a) + 1 / 2 f″[a] (x - a)^2
```

$$1 - \frac{x^2}{2}$$

```
Plot[{f[x], q[x]}, {x, -π, π}, PlotRange → {-1.5, 1.5}]
```

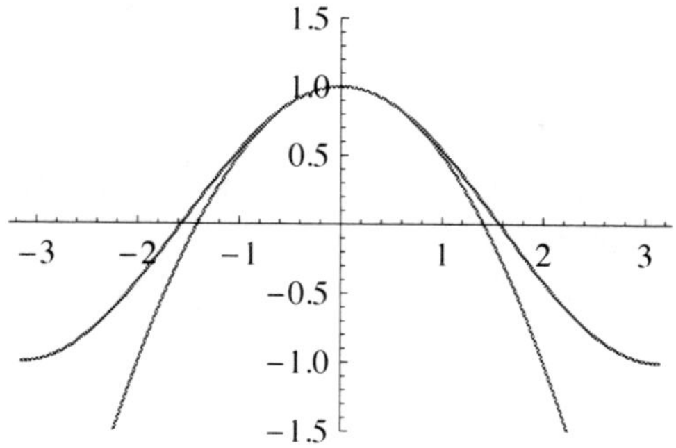

To get an idea of the range of values of x for which $q_0(x)$ gives a good approximation to $\cos(x)$, let's create a plot that shows both graphs on the interval $[-1, 1]$ and a graph of the error $|\cos(x) - q_0(x)|$ on the same interval.

```
GraphicsRow[{Plot[{f[x], q[x]}, {x, -1, 1}],
  Plot[Abs[f[x] - q[x]], {x, -1, 1}, PlotRange → All]}]
```

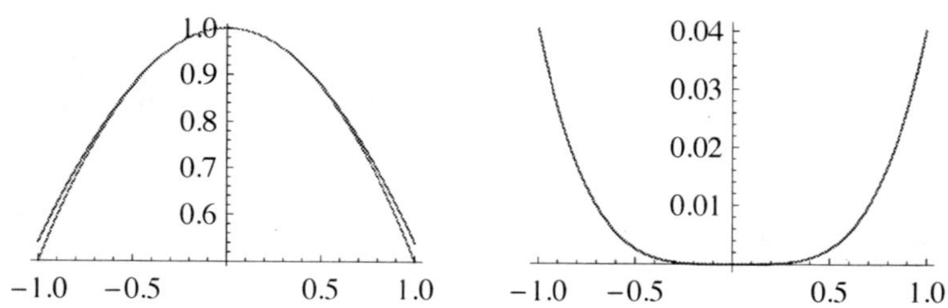

So it appears that $|f(x) - q_0(x)| \lesssim 0.04$ whenever $|x| \leq 1$. Finally, let's make a similar plot for the interval $[-.5, .5]$.

```
GraphicsRow[{Plot[{f[x], q[x]}, {x, -.5, .5}],
  Plot[Abs[f[x] - q[x]], {x, -.5, .5}, PlotRange → All]}]
```

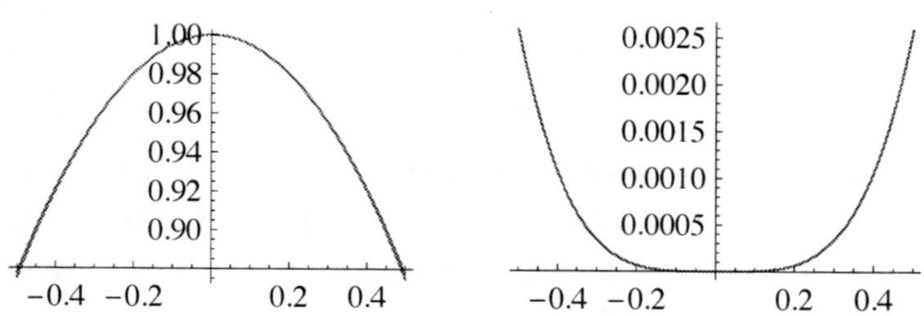

It appears that $|f(x) - q_0(x)| \lesssim 0.0025$ whenever $|x| \leq 0.5$. So we see that the quadratic approximation does a reasonably good job of approximating $\cos x$ over a fairly wide range of values of x.

◆ Exercises

1. a) Find the linear approximation to $f(x) = \sqrt[3]{x}$ at $x = 8$. Then use it to approximate $\sqrt[3]{9}$. Compare the approximation with the correct value.

   b) Find the quadratic approximation to $f(x) = \sqrt[3]{x}$ at $x = 8$. Then use it to approximate $\sqrt[3]{9}$. Compare the approximation with the correct value.

For each function in exercises 2-5, graph the function along with both its linear and quadratic approximations at the specified point.

2. $f(x) = e^{-x}$ at $x = 0$
3. $f(x) = (1 - x)^{-1}$ at $x = 0$
4. $f(x) = \cos x$ at $x = \pi/6$
5. $f(x) = x^3/3$ at $x = 1$
6. Find the equation of the parabola that best approximates the circle $x^2 + y^2 = R^2$ at $(0, -R)$.
7. Use the quadratic approximation to $\cos x$ at $x = 0$ to derive an approximate formula for the solution of $\cos x = kx$ in terms of k. For what values of k does the formula give a reasonably accurate result? Base your answer on the graphs of $y = \cos x$, $y = kx$, and the quadratic approximation.
8. By analogy with the derivation of the quadratic approximation, derive the *cubic* approximation to a function f about $x = a$. Then find the cubic approximation to $f(x) = \sin x$ at $x = 0$.

3.4 Newton's Method

This section is about solving equations, or equivalently, finding zeros of functions. Any equation in one variable can be written in the form

$$f(x) = 0.$$

For example, the equation $x^2 = 5$ is equivalent to $f(x) = 0$ where $f(x) = x^2 - 5$.

The idea behind Newton's method is that an approximation x_0 to a solution of $f(x) = 0$ can be improved by solving the *linearized problem*

$$f(x_0) + f'(x_0)(x - x_0) = 0$$

for a new approximation x_1. This amounts to finding the point x_1 where the tangent line to the graph of f at $(x_1, f(x_1))$ crosses the x-axis. The repetition of this process, generating a sequence of successively improved approximations, is known as Newton's method. From a current approximation x_k we solve the linearized problem

$$f(x_k) + f'(x_k)(x - x_k) = 0$$

to obtain a new approximation x_{k+1}. The resulting formula for x_{k+1} is

$$x_{k+1} = x_k - \frac{f(x_k)}{f'(x_k)}.$$

Notice that each x_k is computed from the previous one by means of the function

$$newt(x) = x - \frac{f(x)}{f'(x)}.$$

This iterative calculation, which is remarkably efficient at quickly producing highly accurate solutions, can be carried out easily in *Mathematica* with the help of the `NestList` function.

▪ Example 3.4.1

It it not uncommon to encounter an equation whose solution(s) cannot be expressed in any exact form and so must be approximated numerically. One such equation is

$$x^2 = \cos \pi x.$$

To view this as a problem of finding the zeros of a function, let

```
f[x_] := x^2 - Cos[π x]
```

From the graph of this function, we see that there are two solutions, each of which is the negative of the other. So let's find the positive solution, which apparently lies in the interval [.4, .5].

```
Plot[f[x], {x, -2, 2}]
```

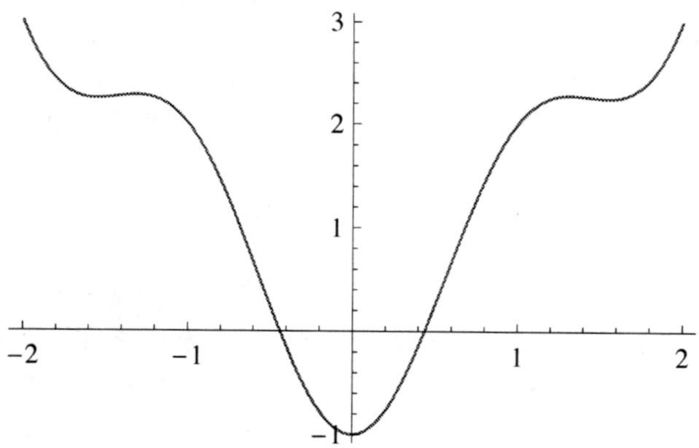

The function that generates the Newton sequence in this case is

```
newt[x_] = x - f[x]/f'[x]
```

$$x - \frac{x^2 - \text{Cos}[\pi x]}{2 x + \pi \text{Sin}[\pi x]}$$

To compute six iterates beginning with $x_0 = .5$, we enter

```
NestList[newt, .5, 6]
```

{0.5, 0.439637, 0.438431, 0.438431, 0.438431, 0.438431, 0.438431}

Notice that after only two iterations, the approximation is accurate to six decimal places. In order to view sixteen decimal digits of each iterate in `TableForm`, enter

```
NumberForm[% // TableForm, 16]
```

```
0.5
0.4396367482486941
0.4384314898117582
0.4384307794817676
0.4384307794815194
0.4384307794815193
0.4384307794815194
```

Notice that after only four iterations, the approximation is accurate to all sixteen places.

⚠ **Warning!** When implementing iterative procedures such as the one above, be sure to force *Mathematica* to do numerical, rather than symbolic, calculations. Notice what happens when we repeat just two steps of the above calculation with the starting point given as 1/2 instead of 0.5.

```
NestList[newt, 1 / 2, 2]
```

$$\left\{\frac{1}{2},\ \frac{1}{2}-\frac{1}{4\,(1+\pi)},\ \frac{1}{2}-\frac{1}{4\,(1+\pi)}-\frac{\left(\frac{1}{2}-\frac{1}{4\,(1+\pi)}\right)^2-\text{Cos}\left[\pi\left(\frac{1}{2}-\frac{1}{4\,(1+\pi)}\right)\right]}{2\left(\frac{1}{2}-\frac{1}{4\,(1+\pi)}\right)+\pi\,\text{Sin}\left[\pi\left(\frac{1}{2}-\frac{1}{4\,(1+\pi)}\right)\right]}\right\}$$

The third and fourth iterations would fill several pages. (Try it.) Such calculations, which can easily require enormous amounts of time and memory, are often the cause of difficulties in *Mathematica*.

◆ Exercises

1. Find and read the description of `NestList` in the *Mathematica* Documentation Center.

In Exercises 2-7, plot the given function and determine crude initial approximations to each of the zeros of the function. Then apply Newton's method to find each zero, accurate to at least six decimal places.

2. $f(x) = x^3 - x^2 - 2x + 1$
3. $f(x) = 2x^5 - 5x - 1$
4. $f(x) = x^2 e^{-x/2} - 1$
5. $f(x) = x - 2.95 \cos x$
6. $f(x) = x^2 - \tan^{-1} x$
7. $f(x) = e^x - x$
8. Find all solutions of $\sin x^2 = \sin^2 x$ in the interval $[0, \pi]$.
9. The equation $3 + \ln x = \sqrt{x}$ has two solutions. Find them both.

Exercises 10-12 demonstrate how Newton's method may perform poorly or fail to find a solution at all in certain circumstances.

10. The only positive zero of $f(x) = x^3 - 3x^2 + 4$ is $x = 2$. Try to find this by Newton's method, starting with an initial approximation $x_0 = 2.1$. Then graph the function. To what property of the function might we attribute the performance of Newton's method on this problem?

11. The function $f(x) = \tan x - x$ has a zero in the interval $29 < x < 30$. Try to find it with Newton's method, beginning with an initial approximation of:

a) $x_0 = 29.75$ b) $x_0 = 29.85$ c) $x_0 = 29.80$

12. The only positive zero of the cubic polynomial $f(x) = x^3 - 5x$ is $\sqrt{5}$. Try to find this with Newton's method, starting with an initial approximation $x_0 = 1$. Repeat with $x_0 = 1.00001$, 1.000001, 1.0000001, and 1.00000001.

3.5 Optimization

One of the most important applications of calculus is optimization. Optimization is about finding maximum and minimum values (i.e., *maxima* and *minima*) of functions. Maxima and minima collectively are referred to as extreme values, or *extrema*. In this section we will look at some general issues related to finding extrema of functions. The next section will be devoted to applied problems.

■ Local Extrema

A function can have numerous local maxima and local minima. The values of x where these extrema occur are among the *critical numbers* of the function, at which either $f'(x) = 0$ or $f'(x)$ fails to exist.

▪ Example 3.5.1

Consider the polynomial

```
f[x_] := x^5 - 4 x^4 - x^3 + 16 x^2 - 12 x
```

The graph of this function indicates two local minima and two local maxima.

```
Plot[f[x], {x, -2.2, 3.3}]
```

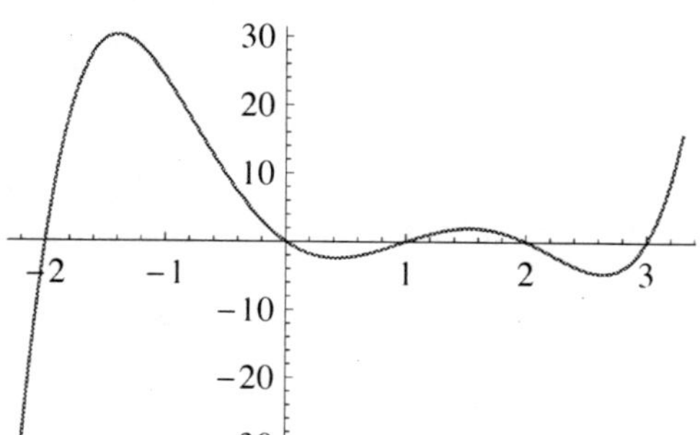

The values of x where these extrema occur can be located by finding the critical numbers of the function, which, since $f'(x)$ exists for all x, are just the zeros of the derivative:

```
critnos = x /. NSolve[f'[x] == 0, x]
{-1.396, 0.425314, 1.53174, 2.63895}
```

The corresponding extrema are

```
f[critnos]
```

```
{30.1602, -2.4034, 1.97771, -4.62861}
```

We must be careful at this point to emphasize the fact that although local extrema often occur at critical numbers where $f'(x) = 0$, such a critical number may just as well produce no local extremum at all. A nearly trivial example is $f(x) = x^3$. The derivative, $f'(x) = 3x^2$, is zero at $x = 0$; yet neither a local maximum nor local minimum occurs there.

```
Plot[x^3, {x, -1, 1}, PlotRange → All]
```

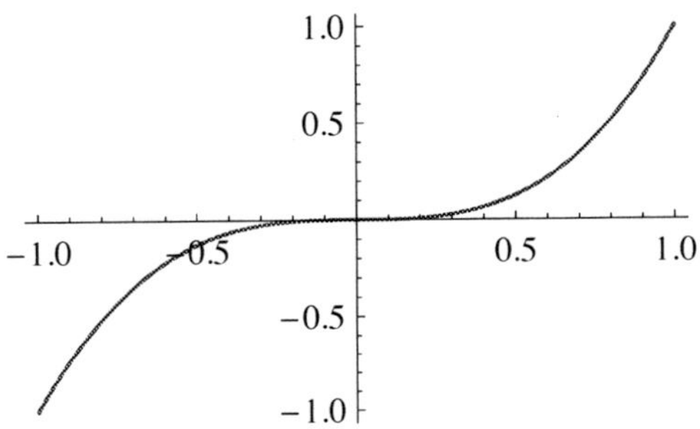

■ The Second Derivative Test

Suppose that c is a critical number of f at which $f'(c) = 0$ and $f''(c)$ exists. The following comprise the **second derivative test**:

- If $f''(c) > 0$, then $f(c)$ is a locally minimum value of f.
- If $f''(c) < 0$, then $f(c)$ is a locally maximum value of f.
- If $f''(c) = 0$, then $f(c)$ could be a local maximum, a local minimum, or neither.

■ Example 3.5.2a

Consider the function

```
f[x_] := 1 - 5 x + 44 x^2 - 152 x^3 + 256 x^4 - 208 x^5 + 64 x^6
```

The derivative of f is

```
f'[x]
```

$$-5 + 88\,x - 456\,x^2 + 1024\,x^3 - 1040\,x^4 + 384\,x^5$$

and the critical numbers are

```
critnos = x /. Solve[f'[x] == 0, x]
```

$$\left\{\frac{1}{2}, \frac{1}{2}, \frac{1}{2}, \frac{1}{48}\left(29 - \sqrt{601}\right), \frac{1}{48}\left(29 + \sqrt{601}\right)\right\}$$

The second derivative of f is

```
f''[x]
```

```
88 - 912 x + 3072 x^2 - 4160 x^3 + 1920 x^4
```

and the values of f'' at the critical numbers are

```
f''[critnos] // N
```

```
{0., 0., 0., 26.3608, 91.1959}
```

Thus the second derivative test tells us that

$$f\left[\frac{1}{48}\left(29-\sqrt{601}\right)\right] \text{ // N}$$

```
0.811036
```

is a local minimum, as is

$$f\left[\frac{1}{48}\left(29+\sqrt{601}\right)\right] \text{ // N}$$

```
-0.378127
```

However, the test is inconclusive at $x = 1/2$. The function's graph reveals that

```
f[.5]
```

```
1.
```

is in fact a local maximum of f.

```
Plot[f[x], {x, -.15, 1.3}]
```

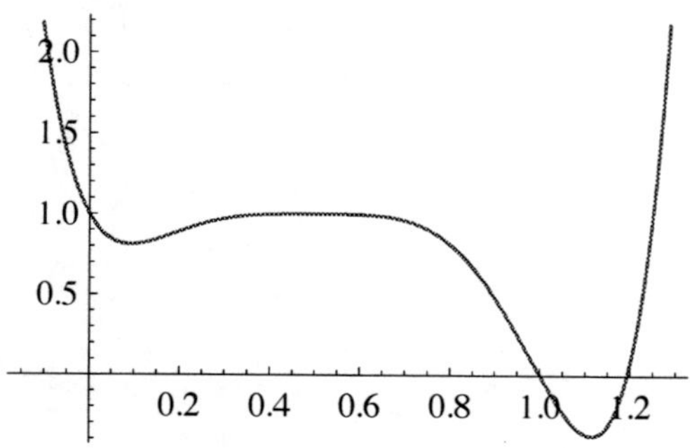

▪ Example 3.5.2b

Consider the function

```
f[x_] := 1 + 5 x - 34 x^2 + 84 x^3 - 88 x^4 + 32 x^5
```

The derivative of f is

```
f'[x]
```

```
5 - 68 x + 252 x^2 - 352 x^3 + 160 x^4
```

and the critical numbers are

```
critnos = x /. Solve[f'[x] == 0, x]
```

$$\left\{\frac{1}{2}, \frac{1}{2}, \frac{1}{20}\left(12 - \sqrt{94}\right), \frac{1}{20}\left(12 + \sqrt{94}\right)\right\}$$

The second derivative of f is

```
f''[x]
```

$$-68 + 504\,x - 1056\,x^2 + 640\,x^3$$

and the values of f'' at the critical numbers are

```
f''[critnos] // N
```

```
{0., 0., -22.9658, 53.0458}
```

Thus the second derivative test tells us that

$$f\left[\frac{1}{20}\left(12 - \sqrt{94}\right)\right] \; // \; N$$

```
1.23836
```

is a locally maximum value of f and that

$$f\left[\frac{1}{20}\left(12 + \sqrt{94}\right)\right] \; // \; N$$

```
-0.146916
```

is a local minimum. Again, the test is inconclusive at $x = 1/2$. A plot of the function's graph reveals that f has neither a local minimum nor a local maximum at $x = 1/2$. (Indeed, $(1/2,\ f(1/2))$ is an inflection point.)

```
Plot[f[x], {x, -.15, 1.3}]
```

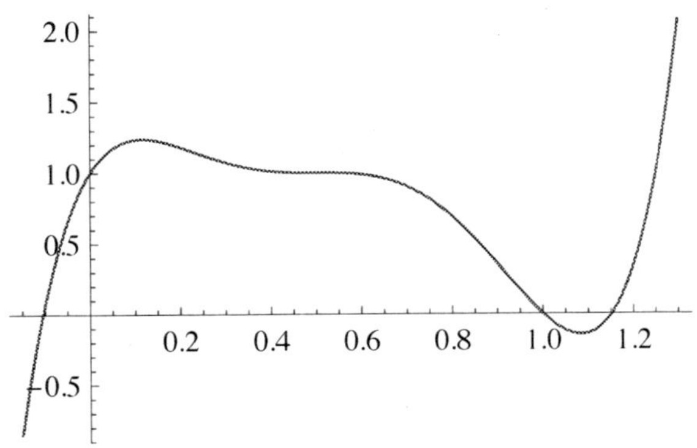

■ Using `FindMinimum` and `FindMaximum`

Mathematica's `FindMinimum` and `FindMaximum` functions use numerical search algorithms to locate extrema. The two minima of the function in Example 3.5.1 are found as follows. The second argument to `FindMinimum` is a list containing the variable and an initial approximation to the location of the sought minimum.

```
f[x_] := x^5 - 4 x^4 - x^3 + 16 x^2 - 12 x
FindMinimum[f[x], {x, .5}]
{-2.4034, {x → 0.425314}}
FindMinimum[f[x], {x, 2.5}]
{-4.62861, {x → 2.63895}}
```

Notice that the result is a list containing the found minimum value and its location in the form of a rule.

To find the maxima of $f(x)$ we use `FindMaximum` similarly.

```
FindMaximum[f[x], {x, -1}]
{30.1602, {x → -1.396}}
FindMaximum[f[x], {x, 1.5}]
{1.97771, {x → 1.53174}}
```

■ Other Kinds of Critical Numbers

In addition to the zeros of f', values of x (in the domain of f) where $f'(x)$ does not exist are also critical numbers of f.

■ Example 3.5.3

A nice example is provided by $f(x) = \sqrt[3]{|x|-2}$. As usual when dealing with the cube-root function, we need to be careful to define the function so that it gives real values.

```
f[x_] := If[-2 < x < 2, -∛(2 - Abs[x]), ∛(Abs[x] - 2)]
Plot[f[x], {x, -6, 6}]
```

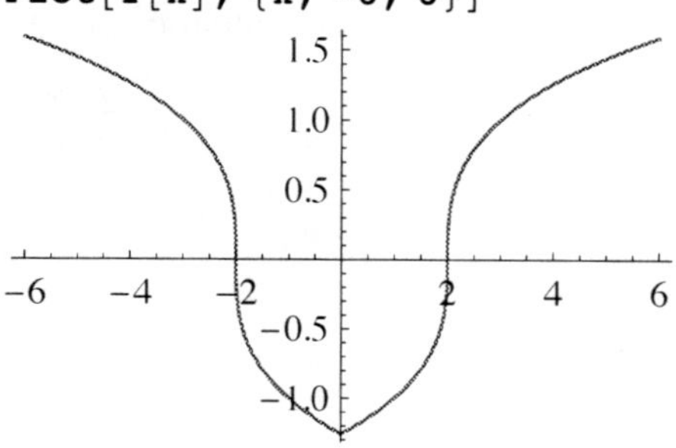

The derivative of this function fails to exist at $x = 0, \pm 2$ and is nowhere zero. Note that the tangent line is vertical at $x = \pm 2$ and that there is no unique tangent line at $(0, -2^{1/3})$ since the slope of the graph jumps instantaneously from a negative value to a positive value there.

Also note that this function has a local (and global) minimum at $x = 0$, the place at which the derivative does not exist.

`FindMinimum` can find minima that occur at such critical numbers. Let's first try searching near $x = 1$.

```
FindMinimum[f[x], {x, 1}]
```

FindMinimum::lstol :
The line search decreased the step size to within tolerance specified by AccuracyGoal and PrecisionGoal but was unable to find a sufficient decrease in the function. You may need more than MachinePrecision digits of working precision to meet these tolerances. ≫

$\{-1.25992, \{x \to -2.61265 \times 10^{-9}\}\}$

The resulting message tells us that we should not put too much confidence in the result. If we provide *two* starting values rather than one, a different algorithm is employed, and no warning is issued.

```
FindMinimum[f[x], {x, -1, 1}]
```

$\{-1.25992, \{x \to 2.8634 \times 10^{-11}\}\}$

This indicates that the minimum indeed occurs at $x = 0$.

Absolute Extrema on Closed Intervals

One of the most important theorems in calculus states that

> *a continuous function on a closed, bounded interval attains both an absolute minimum and an absolute maximum value on that interval.*

Moreover, each of these absolute extrema either is a local extremum occuring at a critical number in the interior of the interval or else occurs at one of the endpoints of the interval.

Example 3.5.4

Consider the function

```
f[x_] := x^3 - x
```

on the closed interval $[-1, 3/2]$. It is clear from the graph that the (absolute) maximum value of f occurs at the right endpoint $x = 3/2$ and that the (absolute) minimum of f occurs at a critical number somewhere near 0.6.

```
Plot[f[x], {x, -1, 3/2}]
```

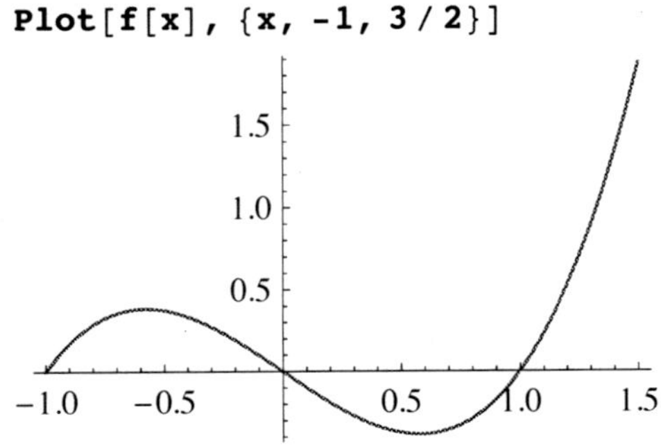

The critical numbers of f are

```
Solve[f'[x] == 0, x]
```

$$\left\{\left\{x \to -\frac{1}{\sqrt{3}}\right\}, \left\{x \to \frac{1}{\sqrt{3}}\right\}\right\}$$

Thus, the absolute minimum and maximum values of f on $[-1, 3/2]$ are

```
{f[1/√3], f[3/2]}
```

$$\left\{-\frac{2}{3\sqrt{3}}, \frac{15}{8}\right\}$$

■ Example 3.5.5

Without the aid of a graph, find the absolute maximum and minimum values of f on the interval [0, 4], *where*

```
f[x_] := (3 x + 1) (2 x - 1) (3 x - 5) (x - 3) (2 x - 9) (x - 5);
Expand[f[x]]
```

$$-675 + 240\, x + 4534\, x^2 - 5836\, x^3 + 2665\, x^4 - 516\, x^5 + 36\, x^6$$

First we need to find the critical numbers of f.

```
critnos = x /. NSolve[f'[x] == 0, x]
```

```
{-0.0252197, 1.03028, 2.34919, 3.80039, 4.78981}
```

The next step is to discard any critical numbers that are not inside the interval.

```
critnos = Select[critnos, Function[x, 0 < x < 4]]
```

```
{1.03028, 2.34919, 3.80039}
```

Then we add the endpoints of the interval to the list.

```
candidates = Union[critnos, {0, 4}]
```

```
{0, 1.03028, 2.34919, 3.80039, 4}
```

Finally, we compute the value of f at each number in the list and find the minimum and maximum values from among the resulting list.

```
vals = f[candidates]
```

```
{-675, 449.447, -452.251, 703.955, 637}
```

```
{Min[vals], Max[vals]}
```

```
{-675, 703.955}
```

Thus we see that the minimum value of f on $[0, 4]$ is $f(0) = -675$ and the maximum is $f(3.80039) = 703.955$. Now let's peek at the graph.

```
Plot[f[x], {x, 0, 4}]
```

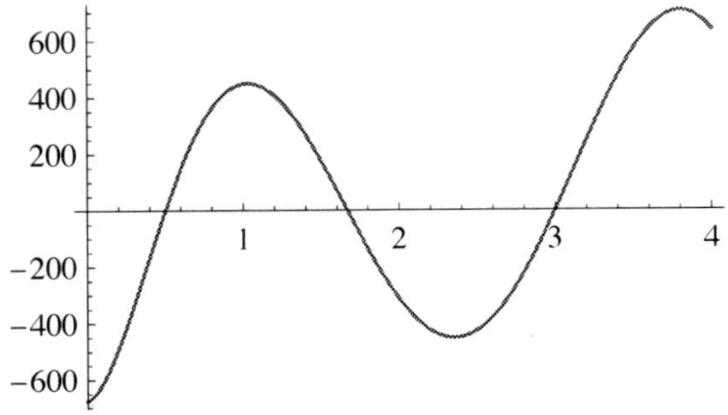

◆ Exercises

1. Find and read the descriptions of `NSolve` and `FindMinimum` in the *Mathematica* Documentation Center.

For the functions in Exercises 2-5,

a) find all critical points of the function;

b) find each local extremum and the value of x at which it occurs.

2. $f(x) = x^5 - x^3$
3. $f(x) = x^3\, e^{-x^2}$
4. $f(x) = x^{1/3}(x - 1)$
5. $f(x) = \left| x^3 - x^2 \right|$

In Exercises 6-9, find the (absolute) maximum and minimum values of the function on the specified closed interval.

6. $f(x) = \ln x$ on $[1, 3]$
7. $f(x) = \left| 3\,x^3 - 2\,x^2 \right|$ on $[0, 1]$
8. $f(x) = x\,(x-1)\,(x-2)$ on $[0, 2]$
9. $f(x) = e^{-x} \sin 3\,x$ on $[0, 5]$
10. Find the absolute minimum value of

$$f(x) = \frac{x^4 - x^3 + 5\,x + 3}{x^2} \text{ on } (0, \infty).$$

Justify your answer *two ways*.

3.6 Applied Optimization Problems

Optimization is such an important topic because of its applications. Many applications, such as the one in the following example, come from business and manufacturing.

■ Example 3.6.1

An aquarium is to be constructed to hold 20 cubic feet of water. The two ends of the aquarium are to be square, and the aquarium has no top. The glass used for the four sides costs \$0.50 per square foot, while cheaper glass used for the bottom costs \$0.35 per square foot. Glue and rubber caulking to fasten and seal the joints between pieces of glass costs \$0.10 per foot. Finally, framing around the bottom and top perimeters costs \$0.05 per foot. Find the dimensions of the aquarium that minimize the total material cost.

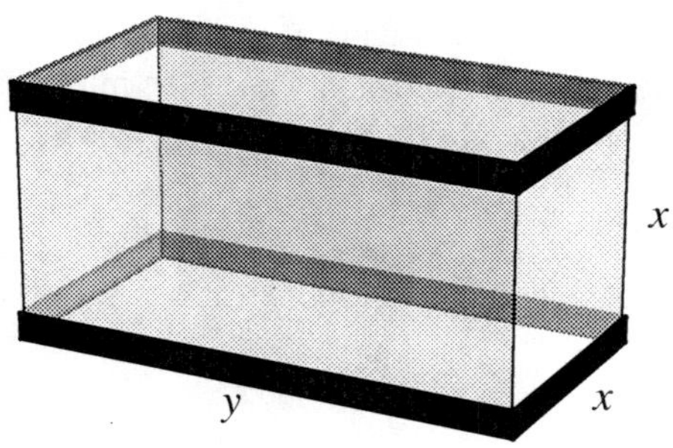

Let x and y be the dimensions indicated in the figure. We will first express the total material cost in terms of x and y.

Cost of the glass for the sides and bottom: $0.50\left(2x^2 + 2\,x\,y\right) + .35\,x\,y$.

Cost of the glue and caulking for the joints: $0.10\,(6x + 2y)$.

Cost of the framing : $0.05\,(2\,(2x + 2y))$.

Putting all this together gives us the total cost as a function of x and y:

```
costxy[x_, y_] =
  .5 (2 x^2 + 2 x y) + .35 x y + .1 (6 x + 2 y) + .05 (2 (2 x + 2 y));
```

Now, because the volume is to be 20 cubic feet, we require that $x^2\,y = 20$. Substitution of $y = 20/x^2$ into the cost function gives the cost as a function of x alone:

```
cost[x_] = costxy[x, 20/x^2] // Simplify
```

$$\frac{8. + 27.\,x + 0.8\,x^3 + 1.\,x^4}{x^2}$$

```
Plot[cost[x], {x, 0, 10}, PlotRange → {0, 100}]
```

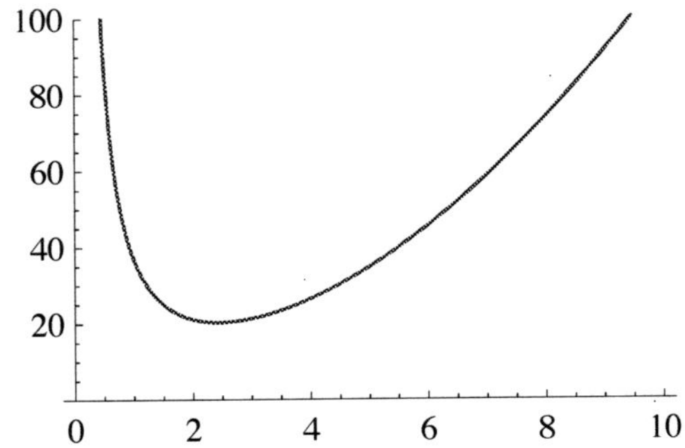

The graph indicates that this cost function has only one critical number of interest:

```
FindRoot[cost′[x], {x, 2}]
{x → 2.43386}
```

This is confirmed by `FindMinimum`, which also reports the minimum cost:

```
FindMinimum[cost[x], {x, 2}]
{20.3148, {x → 2.43386}}
```

Thus the dimensions of the most economical design are $x = 2.43$ feet and $y =$

```
20 / 2.43386^2 feet
3.37628 feet
```

with a resulting minimum cost of $20.31.

The following example is a classical geometric problem that is posed in such a way that we cannot use a completely graphical approach, as was possible in the previous example.

■ Example 3.6.2

Find the dimensions of the smallest right circular cone that can contain a sphere of radius ρ.

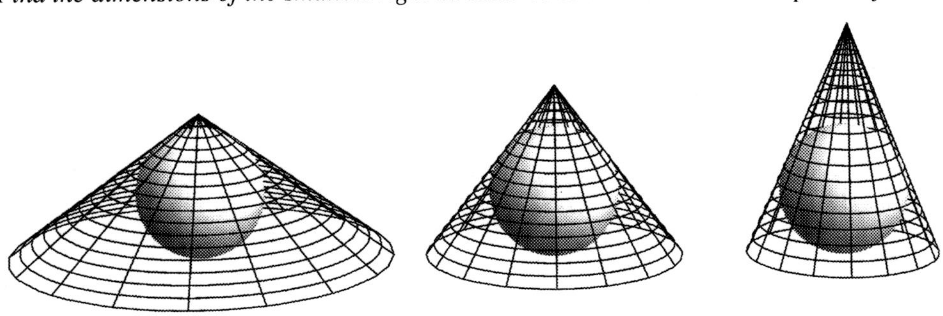

The objective here is to minimize the cone's volume,

```
volrh[r_, h_] := π/3 r^2 h
```

where r and h are the radius and height of the cone, respectively. However, we must first express this volume in terms of one variable. Examining the central cross-section,

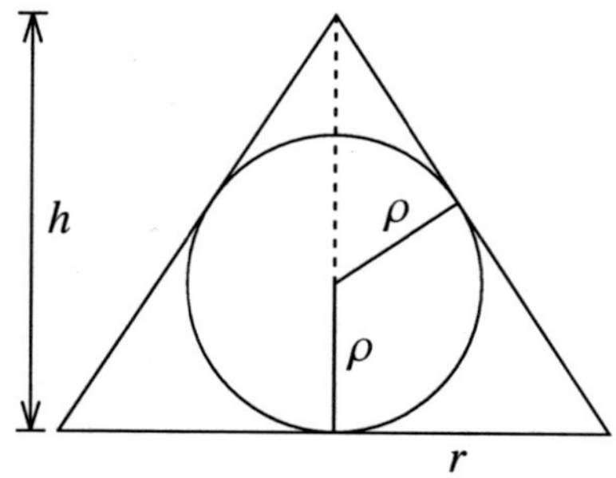

we can use the principle of similar triangles to obtain a relationship between r and h:

```
hrEqn := (h - ρ)/ρ == Sqrt[h^2 + r^2]/r
```

$$\textbf{hrEqn} := \frac{\mathbf{h} - \rho}{\rho} == \frac{\sqrt{\mathbf{h}^2 + \mathbf{r}^2}}{\mathbf{r}}$$

To express the volume as a function of r alone, we'll first solve for h in terms of r:

```
sol = Solve[hrEqn, h] // Flatten
```

$$\left\{h \to 0,\ h \to \frac{2\, r^2\, \rho}{r^2 - \rho^2}\right\}$$

and then substitute the result into the volume formula:

```
h[r_] = h /. sol[[2]];
vol[r_] = volrh[r, h[r]]
```

$$\frac{2\, \pi\, r^4\, \rho}{3\, \left(r^2 - \rho^2\right)}$$

(Note that the natural restriction $r > \rho$ is made clear algebraically by this formula.) Now we compute the derivative with respect to r:

```
vol'[r] // Together // Factor
```

$$\frac{4\, \pi\, r^3\, \rho\, \left(r^2 - 2\, \rho^2\right)}{3\, (r - \rho)^2\, (r + \rho)^2}$$

and find critical numbers:

```
Solve[vol'[r] == 0, r]
```

$$\left\{\{r \to 0\},\ \{r \to 0\},\ \{r \to 0\},\ \left\{r \to -\sqrt{2}\, \rho\right\},\ \left\{r \to \sqrt{2}\, \rho\right\}\right\}$$

Clearly the last of these, $r = \sqrt{2}\,\rho$, is the only critical number of interest. Since the second derivative

```
vol''[Sqrt[2] ρ]
```

$$\frac{32\, \pi\, \rho}{3}$$

is positive, the second derivative test guarantees us that $r = \sqrt{2}\,\rho$ minimizes the volume. Finally, the corresponding cone height is

```
h[√2 ρ]
```

$4\,\rho$

and the minimum volume is

```
vol[√2 ρ]
```

$$\frac{8\,\pi\,\rho^3}{3}$$

To get a better look at what we've done, let's examine the case where $\rho = 1$, in which the optimal radius is $\sqrt{2}$ and the mimum volume is $8\pi/3 \approx 8.38$. This is the graph of the volume versus r:

```
Plot[vol[r] /. ρ → 1, {r, 1, 4}, PlotRange → {0, 40}]
```

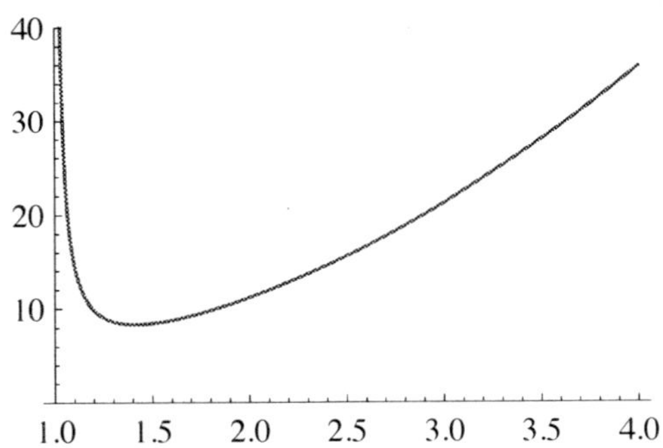

◆ Exercises

1. Rework the aquarium problem ignoring all but the cost of the glass.
2. Rework the aquarium problem if the framing around the bottom and top edges costs \$.50/ft.
3. Find the dimensions of the right circular cone with greatest volume that can be contained in a sphere of radius ρ.
4. Find the dimensions of the right circular cylinder with greatest volume that can be contained in a sphere of radius ρ.
5. Find the area of the largest rectangle (with sides parallel to the coordinate axes) that can be contained in the region in the plane bounded by the graphs of $y = 0$, $y = x^2$, and $x = 1$.
6. Find the area of the largest triangle (with two sides on the coordinate axes) that can be contained in the region in the plane bounded by the graphs of $y = 0$, $y = x^2$, and $x = 1$.
7. Find the point on the graph of $y = 1/x^2$, $x > 0$, that is closest to the origin.
8. Find the minimum surface area of a closed cylinder with a volume of 30 inches^3.

9. Find the maximum volume of a closed cylinder with a surface area of 100 inches^2.

10. Suppose that the rate at which your car burns gasoline is a function of speed (mph) given by

$$f(v) = \left(\frac{v - 37}{37}\right)^3 + 0.01\, v + 1.5 \text{ gallons/hour.}$$

On a 50 mile trip, travelling at constant speed v, the amount of gasoline used will be

$$g(v) = \frac{50\, f(v)}{v} \text{ gallons.}$$

Find the speed v that minimizes this fuel consumption function.

3.7 Antiderivatives

Given a function $f(x)$, a function $F(x)$ such that $F'(x) = f(x)$ is called an **antiderivative** of f. If $F(x)$ is an antiderivative of f, then so is $F(x) + C$ for any constant C, since the derivative of any constant function is zero. In fact, given any antiderivative $F(x)$, *every* antiderivative must have the form $F(x) + C$ for some constant C.

The family of all antiderivatives of f is called the **indefinite integral** of f and is denoted by

$$\int f(x)\, dx$$

So, given any antiderivative $F(x)$ of f, we can describe the indefinite integral by writing

$$\int f(x)\, dx = F(x) + C.$$

For example, since we know that the derivative of x^3 is $3\,x^2$ and the derivative of $\sin x$ is $\cos x$, we have

$$\int 3x^2\, dx = x^3 + C \quad \text{and} \quad \int \cos x\, dx = \sin x + C.$$

Mathematica has the ability to antidifferentiate many types of functions. To compute an antiderivative of a function $f(x)$, we enter either `∫ f[x] dx` or `Integrate[f[x],x]`, as indicated in the following examples. (The integral sign and the special "doublestruck" d are easily obtained by typing [ESC]int[ESC] and [ESC]dd[ESC]. The expression template ∫ □ d□ is on the BasicInput palette as well.)

```
∫ (x + x^3) dx
```

$$\frac{x^2}{2} + \frac{x^4}{4}$$

$$\textbf{Integrate}\left[\frac{x}{x^2 + x - 2}, x\right]$$

$$\frac{1}{3}\text{Log}[-1 + x] + \frac{2}{3}\text{Log}[2 + x]$$

```
Integrate[x Cos[x], x]
```

```
Cos[x] + x Sin[x]
```

$$\textbf{Integrate}\left[x^2 e^{-x}, x\right]$$

$$e^{-x}\left(-2 - 2x - x^2\right)$$

Notice that a single antiderivative is returned, not the indefinite integral. Other antiderivatives are obtained by adding a constant to the one returned by `Integrate`.

By definition, the derivative of any antiderivative of f is f. This can be illustrated by applying the derivative operator ∂_x to the result of `Integrate`.

$$\textbf{Clear[f]};\ \partial_x \int f[x]\, dx$$

```
f[x]
```

Also, any function f is an antiderivative of its derivative; that is,

$$\int f'[x]\, dx$$

```
f[x]
```

■ Special Functions

Many elementary functions do not have antiderivatives that can be expressed in terms of elementary functions. In some cases, *Mathematica* returns an antiderivative of such a function in terms of certain nonelementary *special functions*. For example, the following yield the special functions known as the *error function* and the *Fresnel sine integral*.

$$\int e^{-x^2}\, dx$$

$$\frac{1}{2}\sqrt{\pi}\ \text{Erf}[x]$$

$$\int \text{Sin}\left[x^2\right] dx$$

$$\sqrt{\frac{\pi}{2}}\ \text{FresnelS}\left[\sqrt{\frac{2}{\pi}}\ x\right]$$

We will discuss special functions further in Section 4.4.

Differential Equations and Initial Values

The simplest type of *differential equation* is an equation of the form

$$y'(t) = g(t)$$

where g is a given continuous function, and y is the "unknown." Such an equation is really just an antidifferentiation problem. Its solutions are given by

$$y(t) = \int g(t)\,dt.$$

When the value of the solution at some $t = t_0$ is specified, the problem becomes an *initial value problem*:

$$y'(t) = g(t), \quad y(t_0) = y_0.$$

Typically, such a problem has a solution consisting of one particular function. Given *any* anti-derivative G of g, the solution is

$$y(t) = G(t) - G(t_0) + y_0,$$

which is easily verified as follows:

$$y'(t) = G'(t) - 0 + 0 = g(t), \text{ and } y(t_0) = G(t_0) - G(t_0) + y_0 = y_0.$$

Example 3.7.1

Find the function y that satisfies

$$y'(t) = t^2 \sin t, \quad y(0) = 1/2.$$

First we use `Integrate` to find an antiderivative:

```
antiD[t_] = ∫t² Sin[t] ⅆt
```

```
-(-2 + t²) Cos[t] + 2 t Sin[t]
```

Since every antiderivative of $t^2 \sin t$ differs from this one by a constant, the solution with the desired value at $t = 0$ is

```
antiD[t] - antiD[0] + 1 / 2
```

$$-\frac{3}{2} - (-2 + t^2)\,\text{Cos}[t] + 2\,t\,\text{Sin}[t]$$

Example 3.7.2

Find the function y that satisfies the second-order *initial value problem*

$$y''(t) = t\,e^{-t}, \; y'(0) = 1, \; y(0) = -1.$$

First we use `Integrate` to find an antiderivative of $t\,e^{-t}$:

```
antiD1[t_] = ∫t ⅇ^-t ⅆt
```

```
ⅇ^-t (-1 - t)
```

This function can differ from $y'(t)$ only by a constant. So the desired expression for $y'(t)$ is

```
yprime[t_] = antiD1[t] - antiD1[0] + 1 // Simplify
2 - e^-t (1 + t)
```

Now we antidifferentiate again

```
antiD2[t_] = ∫yprime[t] dt
-e^-t (-2 - t) + 2 t
```

and adjust by the appropriate constant so that $y(0) = -1$:

```
y[t_] = antiD2[t] - antiD2[0] - 1
-3 - e^-t (-2 - t) + 2 t
```

◆ Exercises

In Exercises 1-6, find the indefinite integral by hand, using basic antidifferentiation rules from your textbook. Then check your work by doing the calculation with *Mathematica*.

1. $\int (x^3 - 6x + 3)\, dx$
2. $\int 5\sqrt{x}\, dx$
3. $\int \frac{1}{7x^2}\, dx$
4. $\int (\cos x - \sin x)\, dx$
5. $\int (x + e^x)\, dx$
6. $\int \sin(\pi x)\, dx$

In Exercises 7-12, find the indefinite integral with *Mathematica*. Then check the result by computing its derivative.

7. $\int x\sqrt{x+1}\, dx$
8. $\int \frac{x+3}{x+1}\, dx$
9. $\int \frac{1}{9 + 4x^2}\, dx$
10. $\int (\cos x - \sin x)\, dx$
11. $\int e^{-x} \cos 2x\, dx$
12. $\int \sin^3 x\, dx$

In Exercises 13-15, find and graph the solution of the initial value problem.

13. $y'(t) = \sin 2t,\ y(0) = 0$
14. $y'(t) = t^2 - 4t,\ y(1) = 2$
15. $y''(t) = \sqrt{t},\ y'(0) = -1,\ y(0) = 1$

4 Integration

The fundamental theme underlying all of calculus is calculation of the limit of successively improved approximations. This idea, applied to the problem of finding the slope of a curve, leads to the definition of the derivative. The same idea, when applied to the problem of finding the area under a curve, leads to the definition of the definite integral. In this chapter we will use *Mathematica* to explore some of the basic concepts of integral calculus.

4.1 Limits of Sums and the Area Under a Curve

In this section we will apply the basic idea of calculus—calculation of the limit of successively improved approximations—to the problem of finding the area under the graph of a nonnegative function.

Consider the problem of finding the area of the region in the plane bounded by the graph of $y = x^2$, the x-axis, and the line $x = 1$. Let's first get a picture of this region by graphing $y = x^2$ and shading the region under the graph between $x = 0$ and $x = 1$. We'll do this with the `Filling` option of `Plot`.

```
Plot[x^2, {x, 0, 1}, Filling → Axis]
```

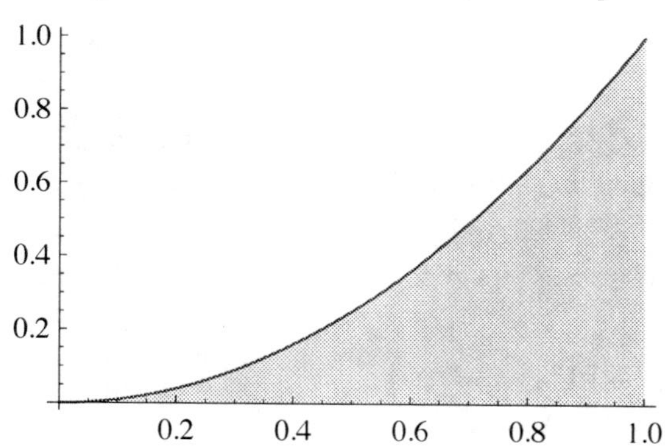

Area Approximations

One approach to approximating the area of the shaded region in the above plot is to approximate the region with a collection of adjacent, non-overlapping rectangles, each of which with height given by some value of the function $f(x) = x^2$.

If we decide to approximate the region with rectangles whose top-left corners touch the curve, then the resulting area approximation amounts to finding the area of a region such as the one shaded below, which shows ten such rectangles, each with width $\Delta x = 0.1$.

(Notice the use of the `Floor` function to graph the "step function" in the picture.)

```
Plot[{x^2, (Floor[10 x] / 10)^2}, {x, 0, 1}, PlotStyle → {Red, Black},
 Filling → {1 → {Axis, Opacity[.1]}, 2 → {Axis, Opacity[.1]}}]
```

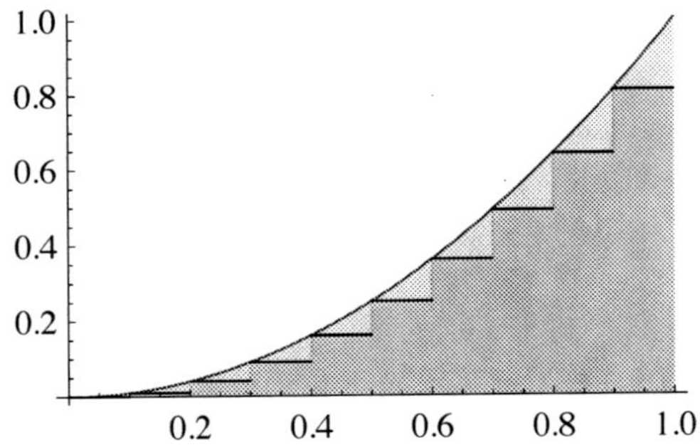

The heights of these rectangles are, respectively, 0^2, $.1^2$, $.2^2$, …, $.9^2$, or

$$\left(\frac{i-1}{10}\right)^2 \text{ for } i = 1, 2, \ldots, 10.$$

Thus the areas of the rectangles are, respectively,

$$\left(\frac{i-1}{10}\right)^2 \frac{1}{10} \text{ for } i = 1, 2, \ldots, 10,$$

and so the sum of their areas is

$$\sum_{i=1}^{10} \left(\frac{i-1}{10}\right)^2 \frac{1}{10} \text{ // N}$$

```
0.285
```

Such an approximation of the area is called a **left-endpoint approximation**. For this particular function, a left-endpoint approximation provides an approximation by *inscribed rectangles*, guaranteed to give an *under*-estimate of the true area under the curve.

Choosing to have the top-right corner of each rectangle touch the curve results in a **right-endpoint approximation**.

```
Plot[{x^2, (Floor[10 x + 1] / 10)^2},
 {x, 0, 1}, PlotStyle → {Red, Black},
 Filling → {2 → {Axis, Opacity[.1]}, 1 → {Axis, Opacity[.1]}}]
```

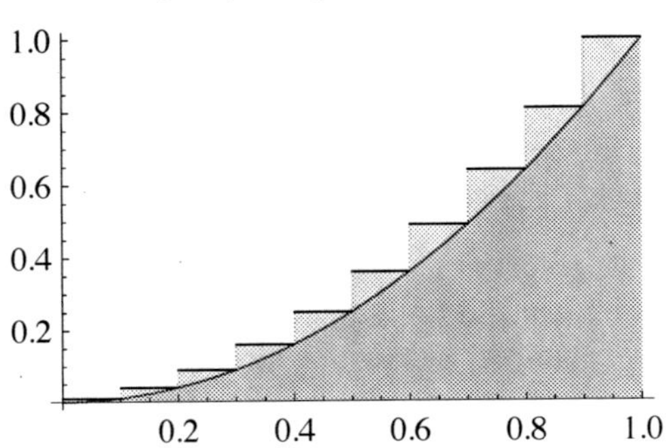

The heights of these rectangles are, respectively, $.1^2,\ .2^2, \ldots, .9^2,\ 1^2$, or

$$\left(\frac{i}{10}\right)^2 \text{ for } i = 1, 2, \ldots, 10.$$

Thus the areas of the rectangles are, respectively,

$$\left(\frac{i}{10}\right)^2 \frac{1}{10} \text{ for } i = 1, 2, \ldots, 10,$$

and so the sum of their areas is

$$\sum_{i=1}^{10} (.1\, i)^2\, .1$$

```
0.385
```

For this particular function, a right-endpoint approximation provides an approximation by *circumscribed rectangles*, guaranteed to give an *over*-estimate of the true area.

An approximation using ten rectangles that is better than either of those above can be had by letting the midpoint of the top of each rectangle touch the curve.

```
Plot[{x^2, ((Floor[10 x] + .5) / 10)^2},
  {x, 0, 1}, PlotStyle → {Red, Black},
  Filling → {2 → {Axis, Opacity[.1]}, 1 → {Axis, Opacity[.1]}}]
```

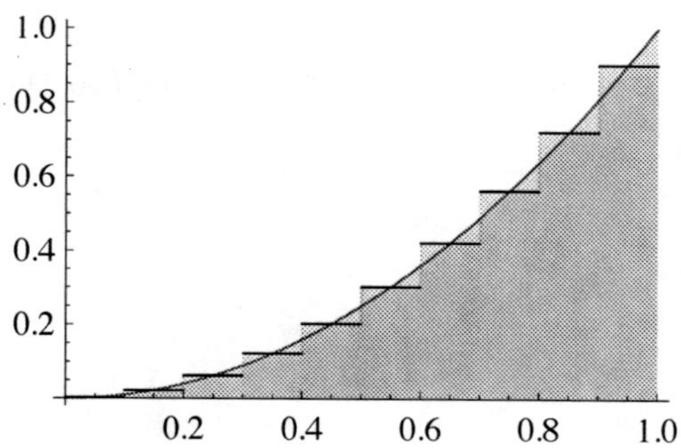

The areas of these rectangles are, respectively, $((i - .5)\cdot 0.1)^2\, 0.1$ for $i = 1, 2, \ldots, 10$, and so the sum of their areas is

$$\sum_{i=1}^{10} \left(\frac{i - .5}{10}\right)^2 \frac{1}{10}$$

```
0.3325
```

Such an approximation to the area is called a **midpoint approximation**.

■ To the Limit!

It should be quite obvious that each of the above approximations can be improved by simply using more rectangles. In fact, the exact area can be obtained by taking the limit of any of these types of approximations as the number of rectangles approaches infinity. So let's compute, for simplicity, the right-endpoint approximation with n rectangles, where n may be any positive integer. By analogy with what we saw above, this area is

```
areaR[n_] := Sum[(i / n)^2 1 / n, {i, 1, n}]
```

$$\text{areaR[n\_]} := \sum_{i=1}^{n} (i/n)^2\, 1/n$$

The following list shows the right-endpoint approximations with 10, 20, 30, ..., 100 rectangles.

```
Table[areaR[n] // N, {n, 10, 100, 10}]
```

```
{0.385, 0.35875, 0.350185, 0.345938, 0.3434,
 0.341713, 0.34051, 0.339609, 0.338909, 0.33835}
```

There is a simple *closed form* for the right-endpoint approximation with n rectangles:

```
areaR[n]
```

$$\frac{(1+n)\ (1+2\ n)}{6\ n^2}$$

Such a closed form makes it easy to find the limit of the right-endpoint approximations as n approaches infinity, which gives the *exact area*.

```
Limit[areaR[n], n → ∞]
```

$$\frac{1}{3}$$

In general, the area under the graph of a continuous, nonnegative function $f(x)$ between $x = a$ and $x = b$ can be found by computing

$$\lim_{n\to\infty} \sum_{i=1}^{n} f(a + i\,\Delta x)\,\Delta x,$$

where $\Delta x = (b - a)/n$.

▪ Example 4.1.1

Compute the area under the graph of $y = x^3 - 5x^2 + 8x - 4$ *between* $x = 1$ *and* $x = 2$.

```
f[x_] := x^3 - 5 x^2 + 8 x - 4
Plot[f[x], {x, 1, 2}, Filling → Axis]
```

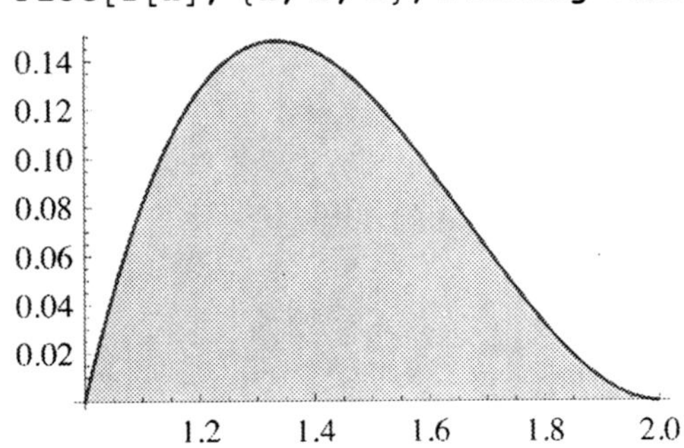

Since $a = 1$ and $\Delta x = \frac{2-1}{n}$, right-endpoint approximations with n rectangles are given by

$$\texttt{areaR[n\_] :=} \sum_{i=1}^{n} f\left[1 + \frac{i}{n}\right] \frac{1}{n}$$

Right-endpoint approximations with 10, 40, 70, and 100 rectangles are

```
Table[areaR[n] // N, {n, 10, 100, 30}] // TableForm
0.0825
0.0832813
0.0833163
0.083325
```

The closed form for right-endpoint approximation with n rectangles is

```
areaR[n] // Simplify
```

$$\frac{-1 + n^2}{12\, n^2}$$

and the exact area is

```
Limit[areaR[n], n → ∞]
```

$$\frac{1}{12}$$

■ Example 4.1.2

Compute the area under the graph of $y = \sin x$ *between* $x = 0$ *and* $x = \pi$.

```
f[x_] := Sin[x]; Plot[f[x], {x, 0, π}, Filling → Axis]
```

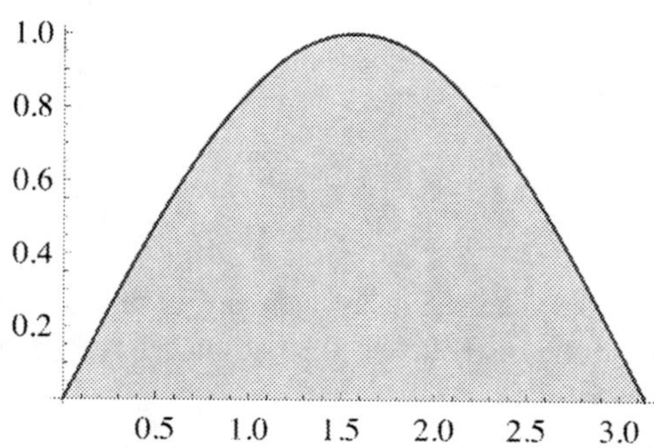

Since $a = 0$ and $\Delta x = \frac{\pi-0}{n}$, right-endpoint approximations with n rectangles are given by

$$\texttt{areaR[n\_] :=} \sum_{i=1}^{n} f\left[0 + \frac{i\,\pi}{n}\right] \frac{\pi}{n}$$

Right-endpoint approximations with 10, 20, 30, …, 100 rectangles are

```
Table[areaR[n] // N, {n, 10, 100, 10}]
{1.98352, 1.99589, 1.99817, 1.99897, 1.99934,
 1.99954, 1.99966, 1.99974, 1.9998, 1.99984}
```

The closed form for right-endpoint approximation with n rectangles is

```
Simplify[areaR[n], Assumptions → {n ≥ 1}]
```

$$\frac{\pi \operatorname{Cot}\left[\frac{\pi}{2\,n}\right]}{n}$$

and the exact area is

```
Limit[areaR[n], n → ∞]
```

2

◆ Exercises

In Exercises 1-3, graph the function and shade the area under the curve. Then, using ten rectangles, compute right-endpoint, left-endpoint, and midpoint approximations to the area under the curve.

1. $f(x) = \sqrt{x}$, $0 \le x \le 1$
2. $f(x) = e^x$, $-1 \le x \le 1$
3. $f(x) = \sin^2 x$, $0 \le x \le \pi$

In Exercises 4 and 5, graph the function and compute a closed form for the sum

$$\sum_{i=1}^{n} f(a + i\,\Delta x)\,\Delta x$$

where $\Delta x = (b - a)/n$. Then compute the limit as $n \to \infty$ to obtain the exact area under the curve.

4. $f(x) = 1 - x^2$, $-1 \le x \le 1$
5. $f(x) = x^2(2 - x)^2$, $0 \le x \le 2$

In Exercises 6 and 7, graph the function and compute the sum

$$\sum_{i=1}^{n} f(a + i\,\Delta x)\,\Delta x$$

where $\Delta x = (b - a)/n$, for $n = 10$, 50, and 100. Then give an estimate of the exact area under the curve along with a statement about the accuracy of the estimate in terms of the number of correct decimal places.

6. $f(x) = 1/x$, $1 \le x \le 2$
7. $f(x) = \sin(x^2/\pi)$, $0 \le x \le \pi$

4.2 The Definite Integral

Let f be a continuous function defined on an interval $[a, b]$ (and not necessarily nonnegative), and let $x_0, x_1, \ldots, x_n$ be the $n+1$ equally spaced points

$$x_0 = a,\ \ x_1 = a + \Delta x,\ \ x_2 = a + 2\,\Delta x,\ \ \ldots,\ \ x_n = b,$$

where $\Delta x = (b-a)/n$. Given a collection of points $x_1^*, x_2^*, \ldots, x_n^*$, where x_i^* is *any* point chosen from the interval $[x_{i-1}, x_i]$, the sum

$$R_n = \sum_{i=1}^{n} f\left(x_i^*\right) \Delta x$$

is called a **Riemann sum** for f on $[a, b]$. Thus the right-endpoint, left-endpoint, and midpoint approximations discussed in the previous section are examples of simple Riemann sums, in which each x_i^* is chosen to be x_{i-1}, x_i, or $(x_{i-1} + x_i)/2$, respectively.

We define the **definite** (or **Riemann**) **integral of** f **from** a **to** b by

$$\int_a^b f(x)\,dx = \lim_{n\to\infty} R_n = \lim_{n\to\infty} \sum_{i=1}^{n} f\left(x_i^*\right) \Delta x.$$

This quantity is well defined because the defining limit always exists if f is continuous on $[a, b]$ and always has the same value regardless of how the x_i^*'s are chosen. Thus the definite integral can be computed as the limit of right-endpoint, left-endpoint, or midpoint approximations.

■ Integrals and Areas

Though the definite integral $\int_a^b f(x)\,dx$ does not give the area "under the graph" of f unless f is nonnegative on $[a, b]$, it *can* always be interpreted in terms of the area(s) of the region(s) bounded by the graph and the x-axis. There are essentially three cases to consider. These are described in the following examples.

■ Example 4.2.1

If $f(x) \geq 0$ on $[a, b]$, then the definite integral gives the area under the graph of f between a and b. Consider

$$\int_0^1 \left(2 - x^2\right) dx.$$

Let's first enter the integrand as $f(x)$.

```
f[x_] := 2 - x^2
```

Now let's create a plot that shades the region between the graph and the x-axis between $x = 0$ and $x = 1$:

```
Show[Plot[f[x], {x, 0, 1}, Filling → Axis],
 Plot[f[x], {x, -1.25, 1.25}], PlotRange → All]
```

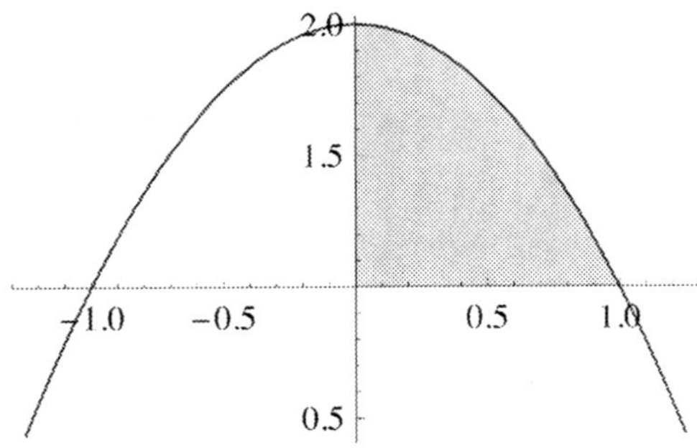

This plot, as well as the fact that $2 - x^2 > 0$ when $0 \le x \le 1$, shows that the region lies entirely above the x-axis. Right-endpoint approximations of the integral with n rectangles are given by

```
a = 0; b = 1;
```

$$\text{intApprox[n\_] := } \sum_{i=1}^{n} f\left[a + \frac{i\ (b - a)}{n}\right] \frac{(b - a)}{n}$$

For instance, 100 rectangles produces the following approximation.

```
intApprox[100] // N
```

```
1.66165
```

A closed form of the approximation with n rectangles is

```
intApprox[n] // Simplify
```

$$-\frac{1+3\,n-10\,n^2}{6\,n^2}$$

The limit as $n \to \infty$ produces the exact value of the integral and the area of the region.

```
Limit[intApprox[n], n → ∞]
```

$$\frac{5}{3}$$

■ Example 4.2.2

If $f(x) \le 0$ on $[a, b]$, then the definite integral gives the *negative of the area* under the graph of f between a and b. Consider

$$\int_0^1 (x^3 - 1)\,dx.$$

Let's first enter the integrand as $f(x)$.

```
f[x_] := x^3 - 1
```

Now we create a plot that shades the region between the graph and the x-axis between $x = 0$ and $x = 1$:

```
Show[Plot[f[x], {x, 0, 1}, Filling → Axis],
 Plot[f[x], {x, -1, 1.5}],
 PlotRange → {{-1, 1.5}, {-1.5, .5}}, AspectRatio → Automatic]
```

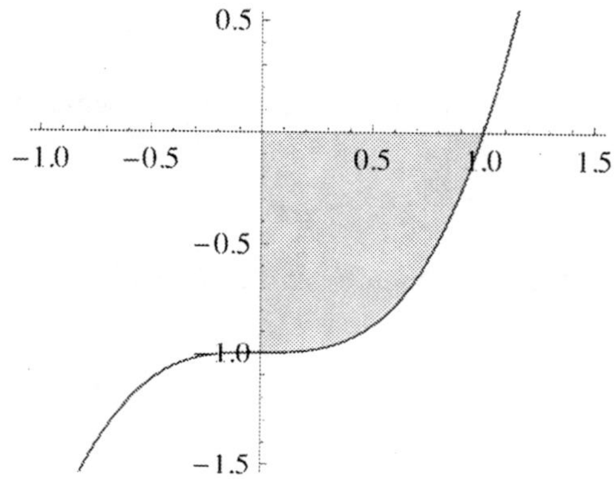

This plot, as well as the fact that $x^3 - 1 \leq 0$ when $0 \leq x \leq 1$, shows that the region lies entirely *below* the x-axis. Right-endpoint approximations of the integral with n rectangles are given by

```
a = 0; b = 1;
```

$$\text{intApprox[n\_]} := \sum_{i=1}^{n} f\left[a + \frac{i\,(b-a)}{n}\right] \frac{(b-a)}{n}$$

For instance, 100 rectangles produces the following approximation.

```
intApprox[100] // N
```

```
-0.744975
```

A closed form of the approximation with n rectangles is

```
intApprox[n] // Simplify
```

$$-\frac{(-1+n)\ (1+3\,n)}{4\,n^2}$$

The limit as $n \to \infty$ produces the exact value of the integral as well as the *negative* of the area of the shaded region.

```
Limit[intApprox[n], n → ∞]
```

$$-\frac{3}{4}$$

So the area of the shaded region is exactly $3/4$.

▪ Example 4.2.3

If $f(x)$ changes sign one or more times on $[a, b]$, then the definite integral gives the "net area" under the graph of f between a and b, that is, the difference between the area above the x-axis and the area below the x-axis. Consider

$$\int_0^2 (x^2 - 1)\, dx.$$

Let's first enter the integrand as $f(x)$.

```
f[x_] := x^2 - 1
```

Now we create a plot that shades the region between the graph and the x-axis between $x = 0$ and $x = 2$:

```
Show[Plot[f[x], {x, 0, 2}, Filling → Axis],
 Plot[f[x], {x, -1.5, 3}],
 PlotRange → {{-1.5, 3}, {-1, 3.25}}, AspectRatio → Automatic]
```

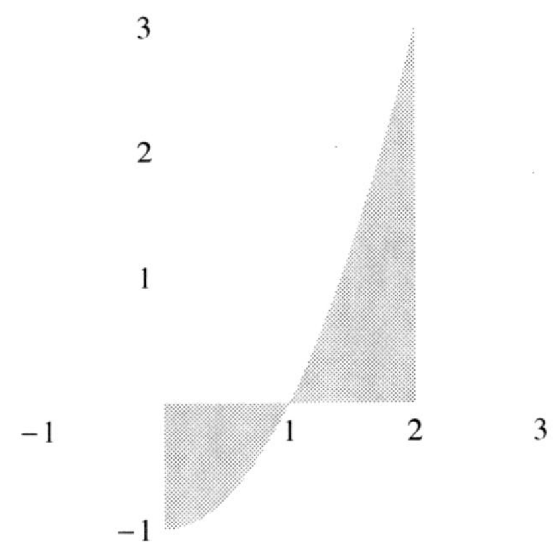

This plot, as well as the fact that $x^2 - 1 < 0$ when $0 \le x < 1$ and $x^2 - 1 > 0$ when $1 < x \le 2$, shows that the region lies partly below and partly above the x-axis. Right-endpoint approximations of the integral with n rectangles are given by

```
a = 0; b = 2;
```

$$\text{intApprox[n\_]} := \sum_{i=1}^{n} f\left[a + \frac{i\,(b-a)}{n}\right] \frac{(b-a)}{n}$$

For instance, 100 rectangles produces the following approximation.

```
intApprox[100] // N
```

```
0.7068
```

A closed form of the approximation with n rectangles is

```
intApprox[n] // Simplify
```

$$\frac{2\,(2 + 6\,n + n^2)}{3\,n^2}$$

The limit as $n \to \infty$ produces the exact value of the integral as well as a "net area" defined to be the area above the x-axis *minus* the area below the x-axis.

```
Limit[intApprox[n], n → ∞]
```

$\frac{2}{3}$

Note that this tells us nothing about the total area of the shaded region. If we wish to determine that total area, we have to compute the areas above and below the x-axis separately and then add them together. In this case, the region below the x-axis corresponds to $0 \le x \le 1$, the region above the x-axis corresponds to $1 \le x \le 2$. So let's compute integrals over those intervals separately. The area between the curve and the x-axis for $0 \le x \le 1$ is

```
a = 0; b = 1;
```

$$\texttt{intApprox[n\_]} := \sum_{i=1}^{n} f\left[a + \frac{i\,(b-a)}{n}\right] \frac{(b-a)}{n}$$

```
area1 = -Limit[intApprox[n], n → ∞]
```

$\frac{2}{3}$

The area between the curve and the x-axis for $1 \le x \le 2$ is

```
a = 1; b = 2;
```

$$\texttt{intApprox[n\_]} := \sum_{i=1}^{n} f\left[a + \frac{i\,(b-a)}{n}\right] \frac{(b-a)}{n}$$

```
area2 = Limit[intApprox[n], n → ∞]
```

$\frac{4}{3}$

So the total area is

```
area1 + area2
```

2

■ Animating the Convergence of Riemann Sums

The following program `integralMovie` creates a sequence of plots that show the area between the graph of a function f between $x = a$ and $x = b$ together with the polygonal region that represents a Riemann sum. It also displays the number of subdivisions and the computed Riemann sum for each plot. In each successive plot the number of subdivisions is doubled, beginning with 2 and stopping with 256. Right-endpoint, left-endpoint, or midpoint approximations can be specified by using 0, .5, or 1 as the fourth argument, respectively.

```
integralMovie[func_, {x_, a_, b_}, {ymin_, ymax_}, sumRule_] :=
  Module[{apprx, f, stepfn, nrects, dx, plot},
   f[t_] = func /. x → t; stepfn[t_, n_] =
    f[a + (Ceiling[n (t - a) / (b - a)] - sumRule) (b - a) / n];
   Animate[dx = N[(b - a) / 2^k];
    apprx = Sum[f[a + (i - sumRule) dx] dx, {i, 1, 2^k}];
    Show[Plot[{f[x], stepfn[x, 2^k]}, {x, a, b}, PlotPoints → 200,
      PlotStyle → {{Red}, {Black}}, Filling → Axis,
      PlotRange → {ymin, ymax}, PlotLabel → {2^k, apprx}]],
    {k, 1, 8, 1}] ];
```

The following is an example of `integralMovie`'s usage. Let's consider the integral

$$\int_0^4 \left(x^3 - 4x^2 - x + 13\right) dx.$$

The following command creates plots in which $0 \le x \le 4$ and $0 \le y \le 13$ and left-endpoint approximations are used.

```
integralMovie[x^3 - 4 x^2 - x + 13, {x, 0, 4}, {0, 13}, .5]
```

◆ Exercises

For each given function and interval in 1-6,

a) create a shaded plot of the graph of the function over the interval;

b) compute the closed form of the right-endpoint approximation to the integral with n rectangles as in the preceding examples;

c) compute the exact value of $\int_a^b f(x)\,dx$ as the limit of right-endpoint approximations;

d) animate the convergence of the right-endpoint approximations with `integralMovie`.

e) interpret the value of the integral as an area, the negative of an area, or a "net area";

1. $f(x) = (x+1)x^3(x-2)^2$, $[a, b] = [-1, 2]$
2. $f(x) = (x+1)x^3(x-2)^2$, $[a, b] = [0, 2]$
3. $f(x) = x\cos x$, $[a, b] = [0, \pi]$
4. $f(x) = x\cos x$, $[a, b] = [0, \pi/2]$
5. $f(x) = x\sin x$, $[a, b] = [0, \pi]$
6. $f(x) = x\sin x$, $[a, b] = [0, \pi/4]$
7. Repeat parts (b) and (c) of 1-6 with left-endpoint and midpoint approximations to verify that the same value of the integral is obtained.

4.3 The Fundamental Theorem of Calculus

Functions Defined by Integrals

Given a continuous function f on an interval $[a, b]$, we can define a function

$$\Phi(x) = \int_a^x f(t)\,dt \quad \text{for } a \le x \le b.$$

A few facts about Φ are easy to notice. First, $\Phi(a) = 0$ and $\Phi(b) = \int_a^b f(x)\,dx$. Also, if $f(x)$ happens to be nonnegative on $[a, b]$, then $\Phi(x)$ gives the area under the graph of f between a and x. It should also be easy to convince yourself that $\Phi(x)$ will be increasing wherever $f(x) > 0$ and decreasing wherever $f(x) < 0$. (Why?)

Example 4.3.1

Let's consider $f(x) = x$ on $[0, 3]$. Then $\Phi(x) = \int_0^x t\,dt$. The following is a plot of both graphs.

```
f[x_] := x; a := 0;
Φ[x_] = Limit[Sum[f[a + i (x - a)/n] (x - a)/n, {i, 1, n}], n → ∞];
Plot[{f[x], Φ[x]}, {x, 0, 3}]
```

$$\Phi[x\_] = \text{Limit}\left[\sum_{i=1}^{n} f\left[a + \frac{i\,(x-a)}{n}\right]\frac{(x-a)}{n},\ n \to \infty\right];$$

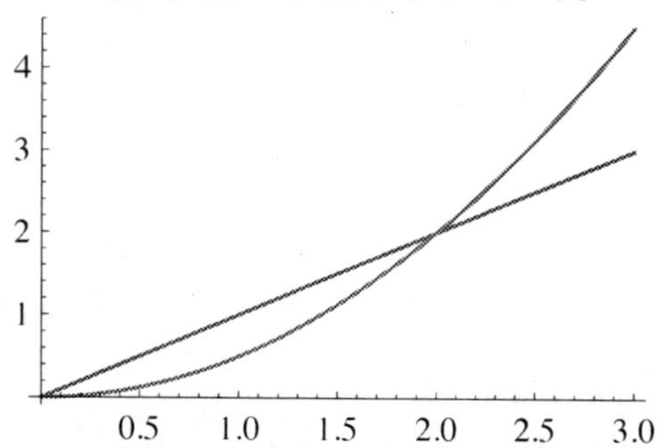

Notice that for each x, $\Phi(x)$ simply gives the area of a triangle with base x and height x.

```
Table[{x, Φ[x]}, {x, 0, 4}] // TableForm
```

0	0
1	$\frac{1}{2}$
2	2
3	$\frac{9}{2}$
4	8

The "closed form" of Φ is simply

```
Φ[x]
```

$$\frac{x^2}{2}$$

Example 4.3.2

Let's consider $f(x) = 2x - x^2$ on [0, 2]. Then $\Phi(x) = \int_0^x (2t - t^2)\,dt$. The following is a plot of both graphs.

```
f[x_] := 2 x - x^2; a := 0;
Φ[x_] = Limit[ Sum[ f[a + i (x - a)/n] (x - a)/n, {i, 1, n}], n → ∞];
Plot[{f[x], Φ[x]}, {x, 0, 3}]
```

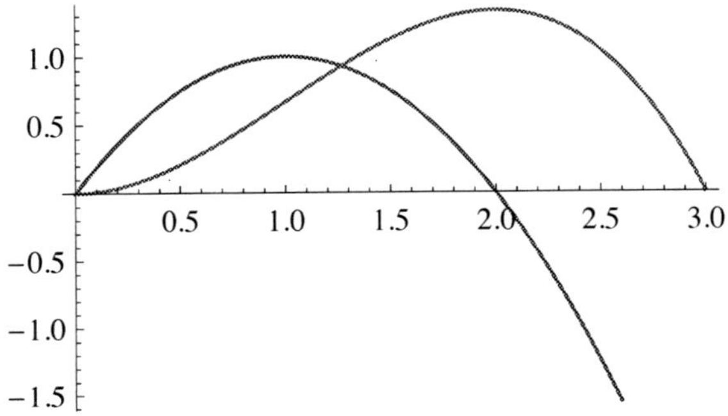

Notice in this plot our previous observation that $\Phi(x)$ is increasing wherever $f(x) > 0$ and decreasing wherever $f(x) < 0$. The closed form of Φ here is

```
Φ[x] // Expand
```

$$x^2 - \frac{x^3}{3}$$

■ Example 4.3.3: Animating the Graph of Φ

The following program creates an animation that plots $\Phi(x) = \int_a^x f(t)\,dt$ along with the region bounded by a curve between a and x, where $a \le x \le b$.

```
areaMovie[func_, {x_, a_, b_}, {ymin_, ymax_}] := Module[{f},
  f[t_] = func /. x → t; int[t_] = Assuming[a < t < b, ∫_a^t f[x] dx];
  If[Head[int[t]] =!= Integrate,
   Animate[Show[
     Plot[f[t], {t, a, b}],
     Plot[f[t], {t, a, x}, Filling → Axis,
      FillingStyle → {Lighter[Red], Lighter[Green]}],
     Plot[int[t], {t, a, x}, PlotStyle → {Blue, Thick}],
     Graphics[{Red, PointSize[Medium], Point[{x, int[x]}]}],
     Graphics[{Gray, Line[{{x, ymin}, {x, ymax}}]}],
     PlotRange → {{a, b}, {ymin, ymax}}], {x, a + .001, b}]]]
```

The following creates an animation of $\Phi(x) = \int_0^x t\,e^{-t/3} \sin t \; dt$ for $0 \le x \le \sqrt{3\pi}$.

```
areaMovie[ x e^(-x/3) Sin[x], {x, 0, 20}, {-1.5, 2}]
```

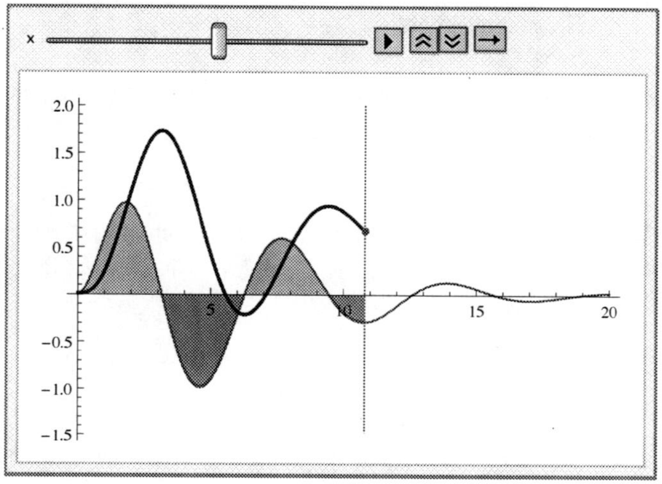

■ The Fundamental Theorem of Calculus

Suppose that f is a continuous function on an interval $[a, b]$ and that

$$\Phi(x) = \int_a^x f(t)\,dt \quad \text{for } a \le x \le b.$$

The *first part* of the **Fundamental Theorem of Calculus** states that $\Phi'(x) = f(x)$ for $a < x < b$; that is:

Every continuous function f on an interval $[a, b]$ has an antiderivative on (a, b), and one such antiderivative is $\Phi(x) = \int_a^x f(t)\,dt$.

■ Example 4.3.4

Consider the function f defined by

```
f[x_] := x + x^2 + x^3
```

The function $\Phi(x) = \int_1^x f(t)\,dt$ has the closed form:

```
a := 1; Φ[x_] = Limit[ Sum[ f[a + i (x-a)/n] (x-a)/n, {i, 1, n}], n → ∞]
```

$$\frac{1}{12}\,(-1 + x)\,\left(13 + 13\,x + 7\,x^2 + 3\,x^3\right)$$

of which it is easy to check that f is the derivative:

```
Φ'[x] // Simplify
```

$$x\,\left(1 + x + x^2\right)$$

This is the *second part* of the Fundamental Theorem of Calculus:

If f is continuous on $[a, b]$ and F is any antiderivative of f on (a, b), then

$$\int_a^b f(x)\,dx = F(b) - F(a).$$

This famous result gives us a very simple and easy way to evaluate definite integrals, *provided* we can find an elementary antiderivative.

Example 4.3.5

Recall from Section 4.1 that `Integrate` can be used to compute antiderivatives. Let's suppose we want to compute the definite integral

$$\int_{-1}^{1} x^2 \sqrt{x+1}\,dx.$$

We first compute an antiderivative,

```
bigF[x_] = ∫x² √(x + 1) dx
```

$$\frac{2}{105}(1+x)^{3/2}\left(8-12x+15x^2\right)$$

and then compute the difference between the antiderivative's values at the endpoints of the interval.

```
bigF[1] - bigF[-1] // Simplify
% // N
```

$$\frac{44\sqrt{2}}{105}$$

0.592623

Since $x^2\sqrt{x+1} \geq 0$ on $[-1, 1]$, the resulting integral gives the area under the graph of $y = x^2\sqrt{x+1}$.

```
Plot[x² √(x + 1), {x, -1, 1}, Filling → Axis]
```

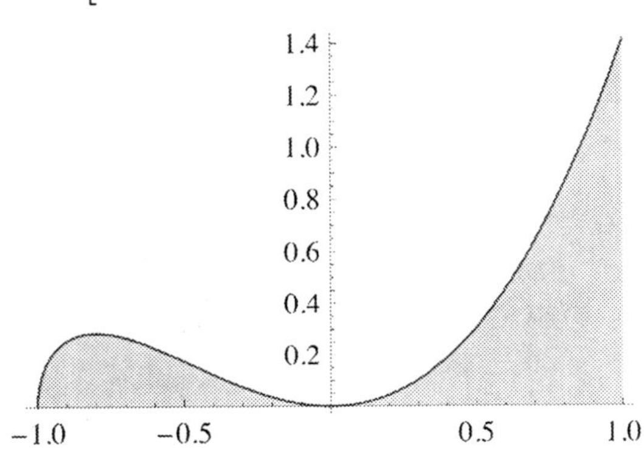

■ Definite Integrals with `Integrate`

The same `Integrate` that computes antiderivatives also computes definite integrals. Using `InputForm`, the integral $\int_1^3 (x^2+1)^{-1}\,dx$, for example, is found by entering

```
Integrate[1/(x^2 + 1), {x, 1, 3}]
% // N
```

$$-\frac{\pi}{4} + \text{ArcTan}[3]$$

```
0.463648
```

Using `StandardForm`, the integral $\int_0^{\pi} \sin 3x \cos 2x\,dx$, for example, is found by entering

$$\int_0^{\pi} \text{Sin}[3\,x]\ \text{Cos}[2\,x]\ dx$$

$$\frac{6}{5}$$

■ Variable Limits of Integration

When given symbolic (variable) limits of integration, `Integrate` attempts to return an answer that is correct for all possible values of that variable, including complex values. Consequently, unless we intervene, `Integrate` can be very slow and will often return a complicated result based upon numerous conditions on the variable(s). The following is a relatively benign example:

$$\int_1^{x} 1/t^2\ dt$$

$$\text{If}\left[\text{Re}[x] \geq 0 \mid\mid \text{Im}[x] \neq 0,\ 1 - \frac{1}{x},\right.$$
$$\left.\text{Integrate}\left[\frac{1}{t^2},\ \{t, 1, x\},\ \text{Assumptions} \to\ !\ (\text{Re}[x] \geq 0 \mid\mid \text{Im}[x] \neq 0)\right]\right]$$

💡 A good way to avoid this is to use `Integrate` *inside of* `Assuming`. In the preceeding integral, we are more than likely only interested in the case where $x > 0$. We make that assumption explicitly as follows:

$$\text{Assuming}\left[x > 0,\ \int_1^{x} 1/t^2\ dt\right]$$

$$\frac{-1 + x}{x}$$

■ Special Functions as Antiderivatives

As we mentioned in Section 4.1, it is quite common for an elementary function not to be the derivative of another elementary function. (See the index in your textbook for the definition of "elementary function.") Two simple examples are $\sin x^2$ and e^{-x^2}.

The antiderivative of $\sin x^2$ is reported by *Mathematica* in terms of `FresnelS`, the *Fresnel sine integral*, which by definition is $\int_0^x \sin(\pi t^2/2)\,dt$.

```
∫Sin[x²] dx
```

$$\sqrt{\frac{\pi}{2}}\ \text{FresnelS}\left[\sqrt{\frac{2}{\pi}}\ x\right]$$

The antiderivative of e^{-x^2} is reported by *Mathematica* in terms of the *error function*, which by definition is $\operatorname{erf}(x) = 2/\sqrt{\pi}\int_0^x e^{-t^2}\,dt$.

```
∫e^(-x²) dx
```

$$\frac{1}{2}\sqrt{\pi}\ \text{Erf}[x]$$

`FresnelS` and `Erf` are each examples of a large collection of special functions that are defined as antiderivatives of elementary functions—elementary functions that are not derivatives of other elementary functions. *Mathematica* is able to compute values of, or plot, these special functions quite efficiently. For example,

```
FresnelS[1] // N
0.438259
Plot[FresnelS[x], {x, -4, 4}]
```

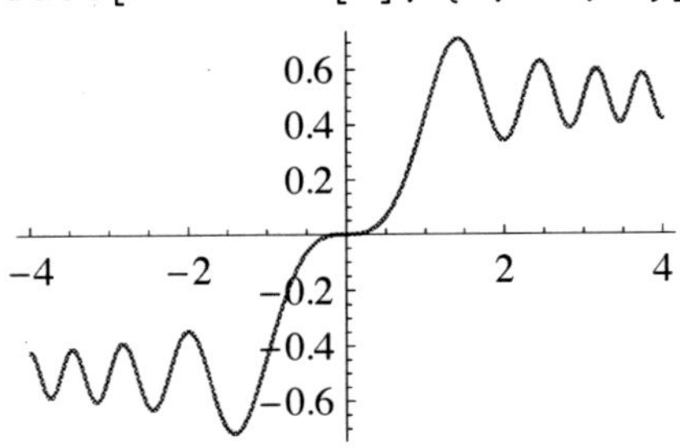

The main point here is that when a continuous function f fails to have an antiderivative that can be expressed in terms of elementary functions, an antiderivative can simply be *defined* as $F(x) = \int_a^x f(t)\,dt$.

■ More on Initial Value Problems

Given a continuous function f, the antiderivative of f that has the value y_0 at $x = a$ is

$$F(x) = y_0 + \int_a^x f(s)\,ds.$$

Thus, the solution $y(t)$ of the initial value problem

$$y'(t) = f(t),\ \ y(a) = y_0$$

is

$$y(t) = y_0 + \int_a^t f(s)\,ds.$$

■ Example 4.3.6

Find the solution of the initial value problem

$$y'(t) = e^{-t} \sin \pi t, \; y(1) = 2$$

The solution is simply

```
y[t_] = 2 + ∫_1^t e^-s Sin[π s] ds
```

$$2 - \frac{e^{-1-t} \left(e^t \pi + e \pi \operatorname{Cos}[\pi t] + e \operatorname{Sin}[\pi t]\right)}{1 + \pi^2}$$

■ Example 4.3.7

Find the solution of the initial value problem

$$y''(t) = t\, e^{-t}, \; y'(0) = -1, y(0) = 1.$$

First we find the first derivative of the solution:

```
yprime[t_] = -1 + ∫_0^t s e^-s ds
```

$-e^{-t} (1 + t)$

Then the solution is found by integrating again:

```
y[t_] = 1 + ∫_0^t yprime[s] ds
```

$-1 + e^{-t} (2 + t)$

◆ Exercises

Use `areaMovie` to create an animation for each of the following.

1. $f(x) = x, \; -1 \le x \le 1$
2. $f(x) = x, \; 0 \le x \le 2$
3. $f(x) = x^2, \; -1 \le x \le 1$
4. $f(x) = x^2 - 1, \; -2 \le x \le 2$
5. $f(x) = |x|, \; -1 \le x \le 1$
6. $f(x) = |x| - 1, \; -2 \le x \le 2$
7. $f(x) = \sin x, \; 0 \le x \le 4\pi$
8. $f(x) = \dfrac{\sin x}{x}, \; 0 \le x \le 6\pi$
9. $f(x) = \sin x^2, \; 0 \le x \le 10$
10. $f(x) = e^{-x^2}, \; -3 \le x \le 3$

In Exercises 11-15, compute the definite integral. Graph the function on the relevant interval and then interpret the value of the integral as an area, the negative of an area, or a "net area."

11. $\displaystyle\int_0^1 20x^3(x-1)\,dx$
12. $\displaystyle\int_0^{2\pi} \sin x\,dx$
13. $\displaystyle\int_0^{2\pi} \sin^2 x\,dx$
14. $\displaystyle\int_{-1}^1 x \sin \pi x\,dx$
15. $\displaystyle\int_{-1}^1 x^2 \sin \pi x\,dx$

Find the solution of each of the initial value problems in 16-19. Plot each solution.

16. $y'(t) = t\,e^{-t},\ y(0) = 0$

17. $y'(t) = t\sqrt{t+1}\,,\ y(0) = 1$

18. $y''(t) = \cos t,\ y'(0) = 1, y(0) = 1$

19. $y''(t) = t - t^2,\ y'(1) = 1, y(1) = 0$

4.4 Numerical Integration

Approximation procedures for definite integrals are of primary importance for at least two reasons. First, many functions do not have antiderivatives that can be expressed in terms of other elementary functions. Also, many important problems arise in which the only thing known about the function to be integrated is a set of values at discrete points.

We have already seen that Riemann sums can be thought of as approximations to an integral. Our goal now is to investigate other numerical approximations that, for a fixed number of subintervals (and therefore a fixed number of function evaluations), provide better approximations.

■ The Midpoint Rule

Before we look at other methods, let's define (for comparison) a *Mathematica* function that computes the usual midpoint approximation. (Right- and left-endpoint approximations are generally quite poor unless a very large number of subintervals are used.)

```
midptApprox[func_, {x_, a_, b_}, n_] :=
 Module[{f, h = (b - a) / n}, f[t_] := func /. x → t; h Sum[f[a + i * h], {i, .5, n - .5}]]
```

For illustration, let's use this to approximate $\int_0^{\pi/2} \sin x \, dx$ with $n = 16$ subintervals. (The exact value of the integral is 1.)

```
midptApprox[Sin[x], {x, 0, π / 2}, 16]
```

```
1.0004
```

■ The Trapezoidal Rule

This procedure is motivated by the idea of using adjacent, non-overlapping trapezoids (rather than rectangles) to approximate the area under a curve. The base of each trapezoid is on the x-axis, the sides are vertical, and each of the top corners touch the graph of the function. Such a trapezoid has an area given by

$$\frac{f(x_{i-1}) + f(x_i)}{2}\,h$$

where x_{i-1} and x_i are the endpoints of the trapezoid's base and $h = x_i - x_{i-1}$ is the width of the base. Summing the areas of n trapezoids produces the integral of the resulting *piecewise-linear approximation* of f, which is the **trapezoidal rule**:

$$\int_a^b f(x)\,dx \approx T_n = \frac{h}{2}\left(f(a) + 2\sum_{i=1}^{n-1} f(x_i) + f(b)\right),$$

where $h = (b-a)/n$ and $x_i = a + ih$. The following is a *Mathematica* function that computes trapezoidal rule approximations.

```
trapez[func_, {x_, a_, b_}, n_] := Module[{f, h = N[(b - a) / n]},
  f[t_] := func /. x → t; 1/2 h (f[a] + 2 Sum[f[a + i h], {i, 1, n-1}] + f[b])]
```

■ Example 4.4.1

For the integral $\int_0^2 \cos^2(\sin x)\,dx$, the midpoint rule with $n = 4$ subintervals gives

```
f[x_] := Cos[Sin[x]]^2;
midptApprox[f[x], {x, 0, 2}, 4]
1.09448
```

The trapezoidal rule approximation with $n = 4$ subintervals is

```
trapez[f[x], {x, 0, 2}, 4]
1.10708
```

The following plot shows the region whose area is given by this approximation, along with the graph of $f(x) = \cos^2(\sin x)$.

```
linapprox = Interpolation[
   Table[{x, f[x]}, {x, 0, 2, .5}], InterpolationOrder → 1];
Show[Plot[{f[x], linapprox[x]}, {x, 0, 2}, PlotStyle → Black,
  Filling → {{2 → {Axis, {Lighter[Yellow, .7]}}},
    {1 → {{2}, {Green, Red}}}}, PlotRange → {0, 1}],
 Graphics[Line /@ ({{#, 0}, {#, f[#]}} &) /@ {.5, 1., 1.5, 2}]]
```

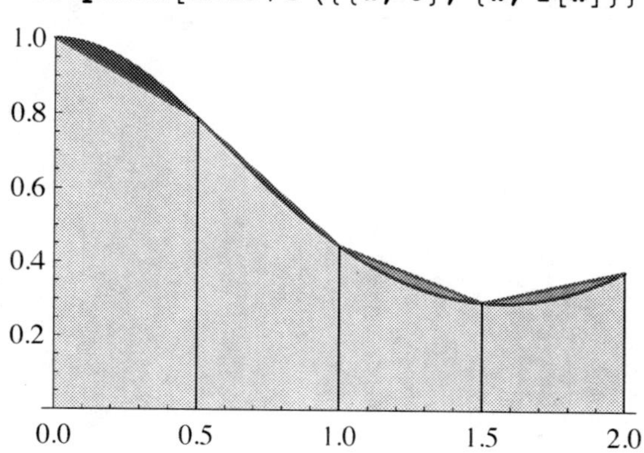

Simpson's Rule

The idea behind Simpson's rule is to approximate the graph of f with a collection of connected parabolic segments, each of which is determined by three points on the curve. Then the integral of f is approximated by the integral of the resulting *piecewise-quadratic approximation* to f. All of this results in **Simpson's rule**:

$$\int_a^b f(x)\,dx \approx S_n = \frac{h}{3}\left(f(a) + 4\,f(x_1) + 2\,f(x_2) + 4\,f(x_3) + \cdots + 4\,f(x_{n-1}) + f(b)\right),$$

where $h = (b-a)/n$ and $x_i = a + ih$. Notice the pattern of the coefficients: 1-4-2-4-2-4···2-4-1, which works only if n is even! Also notice that the alternating 4-2-4-2 pattern can be described by $3 + (-1)^{i-1}$ for $i = 1, 2, \ldots, n-1$.

The following is a *Mathematica* function that computes Simpson's rule approximation. (Notice what happens when n is not even.)

```
simpson[func_, {x_, a_, b_}, n_] := Module[{f, h = (b - a) / n, wt},
   wt[i_] := 3. + (-1)^(i-1); f[t_] := func /. x → t;
  If[EvenQ[n], h/3 (f[a] + Sum[wt[i] f[a + i h], {i, 1, n-1}] + f[b]), "n must be even"]]
```

Example 4.4.2

Consider again the integral $\int_0^2 \cos^2(\sin x)\,dx$. The Simpson's rule approximation with $n = 4$ subintervals is computed as follows.

```
f[x_] := Cos[Sin[x]]^2; simpson[f[x], {x, 0, 2}, 4]
1.09853
```

The following plot shows the region whose area is given by this approximation, along with the graph of $f(x) = \cos^2(\sin x)$.

```
order2 = InterpolationOrder → 2;
quadapprox := If[x ≤ 1,
   Interpolation[Table[{t, f[t]}, {t, 0, 1, .5}], order2],
   Interpolation[Table[{t, f[t]}, {t, 1, 2, .5}], order2] ];
Plot[{f[x], quadapprox[x]}, {x, 0, 2}, PlotRange → {0, 1}, Filling →
  {{2 → {Axis, {Lighter[Yellow, .7]}}}, {1 → {{2}, {Green, Red}}}}]
```

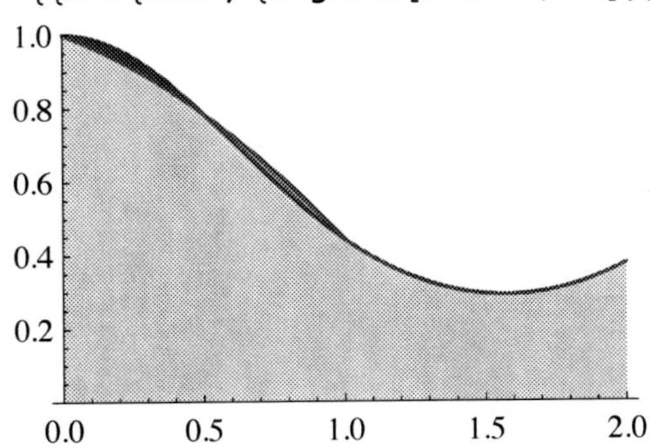

■ Error Bounds

For trapezoidal, midpoint, and Simpson's rule approximations T_n, M_n, and S_n of an integral $I = \int_a^b f(x)\,dx$, errors are given respectively by

$$E_{T_n} = I - T_n,\ E_{M_n} = I - M_n, \text{and } E_{S_n} = I - S_n.$$

Suppose that we have the following bounds on the second and fourth derivatives of f:

$$|f''(x)| \le K \text{ and } \left|f^{(4)}(x)\right| \le C \text{ for all } x \text{ in } [a, b].$$

Then the errors in the trapezoidal, midpoint, and Simpson's rules, respectively, satisfy

$$|E_{T_n}| \le \frac{K\,(b-a)^3}{12\,n^2} = \frac{K\,(b-a)}{12}\,h^2,$$

$$|E_{M_n}| \le \frac{K\,(b-a)^3}{24\,n^2} = \frac{K\,(b-a)}{24}\,h^2,$$

$$|E_{S_n}| \le \frac{C\,(b-a)^5}{180\,n^4} = \frac{C\,(b-a)}{180}\,h^4.$$

■ Example 4.4.3

Consider again the midpoint and trapezoidal rule approximations we computed in Example 4.4.1 for the function $f(x) = \cos^2(\sin x)$ on $[0, 2]$:

```
f[x_] := Cos[Sin[x]]^2
trapez[f[x], {x, 0, 2}, 4]
midptApprox[f[x], {x, 0, 2}, 4]
1.10708
1.09448
```

The second derivative of f is

```
D2f[x_] = ∂x,x f[x] // Simplify
-2 Cos[x]^2 Cos[2 Sin[x]] + Sin[x] Sin[2 Sin[x]]
```

The graph of $|f''(x)|$ makes it clear that

$$|f''(x)| \le 2 \text{ for all } x \text{ in } [0, 2].$$

```
Plot[Abs[D2f[x]], {x, 0, 2}, PlotRange → All]
```

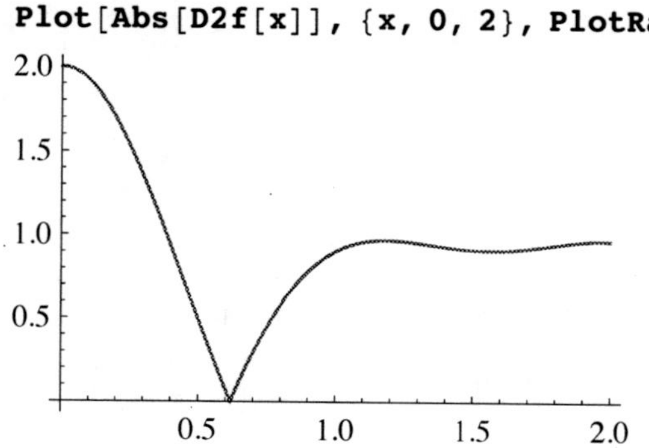

So we use $K = 2$, along with $a = 0,\ b = 2$, and $h = .5$, in the error estimates

$$|E_{T_4}| \le \frac{K(b-a)}{12} h^2 \text{ and } |E_{M_4}| \le \frac{K(b-a)}{24} h^2.$$

This results in the following bound on the trapezoidal rule error:

```
2 (2 - 0)
--------- .5^2
   12
0.0833333
```

and the following bound on the midpoint rule error:

```
2 (2 - 0)
--------- .5^2
   24
0.0416667
```

▪ Example 4.4.4

Consider again the Simpson's rule approximation we computed in Example 4.4.2.

```
f[x_] := Cos[Sin[x]]^2
simpson[f[x], {x, 0, 2}, 4]
1.09853
```

The fourth derivative of f is

```
D4f[x_] = ∂x,x,x,x f[x] // Simplify
8 Cos[x]^4 Cos[2 Sin[x]] +
 8 Cos[x]^2 (Cos[Sin[x]]^2 - Sin[Sin[x]]^2 - 3 Sin[x] Sin[2 Sin[x]]) -
 Sin[x] (3 Sin[x - 2 Sin[x]] + Sin[2 Sin[x]] + 3 Sin[x + 2 Sin[x]])
```

The graph of $|f^{(4)}(x)|$ makes it clear that

$$|f^{(4)}(x)| \le |f^{(4)}(0)| = 16 \text{ for all } x \text{ in } [0, 2].$$

```
Plot[Abs[D4f[x]], {x, 0, 2}]
```

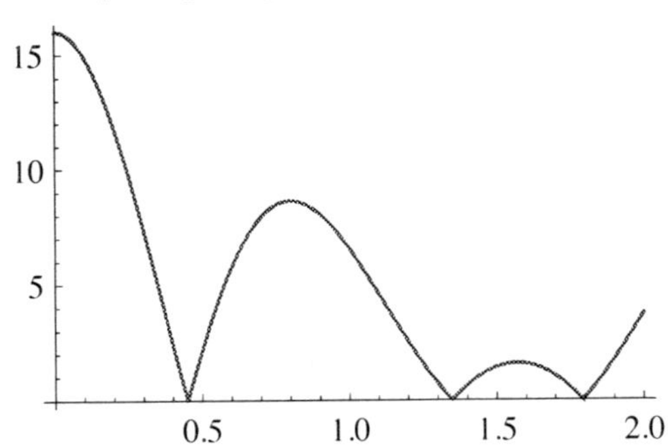

So we use $M = 16$, along with $a = 0, b = 2$, and $h = .5$, in the error estimate

$$|E_{S_4}| \le \frac{M(b-a)}{180} h^4.$$

This results in the the following bound on the error:

```
16 (2 - 0) .5^4 / 180
0.0111111
```

▪ Example 4.4.5

Determine how many subintervals are needed to approximate the integral

$$\int_1^3 e^{-x^2/2}\, dx$$

by the midpoint rule with an error of not more than 0.00005. Then repeat using the trapezoidal rule and then Simpson's rule.

Let's begin by defining the function and then computing its second derivative.

```
f[x_] := e^(-x^2/2)
D2f[x_] = ∂x,x f[x] // Simplify
e^(-x^2/2) (-1 + x^2)
```

Now let's plot the absolute value of f'':

```
Plot[Abs[D2f[x]], {x, 1, 3}]
```

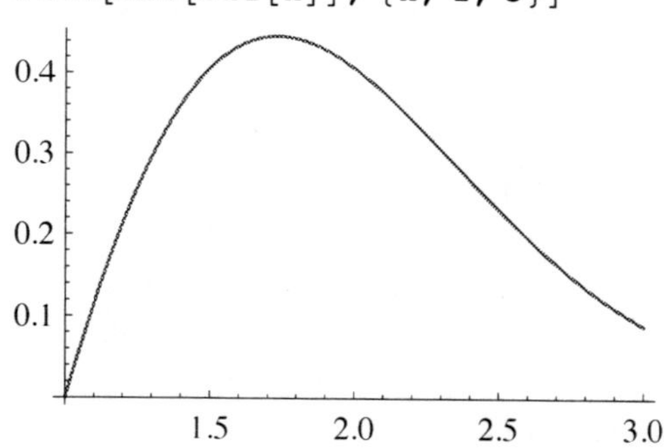

To determine where the maximum value occurs, we'll find the zero of the third derivative that lies near 1.7.

```
FindRoot[Evaluate[∂x D2f[x]], {x, 1.7}]
{x → 1.73205}
```

Now the value of K that we want is

```
bigK = Abs[D2f[x]] /. %
0.44626
```

and the resulting error bound for the midpoint rule with n subintervals is

```
midptErrorBound[n_] = bigK (3 - 1)^3 / (24 n^2)
0.148753 / n^2
```

The following table of values shows that our error bound is less than .00005 when $n \geq 55$.

```
Table[{n, midptErrorBound[n]}, {n, 53, 57}] // TableForm
```

```
53   0.000052956
54   0.0000510128
55   0.0000491747
56   0.0000474341
57   0.0000457844
```

The resulting midpoint approximation is

```
NumberForm[midptApprox[f[x], {x, 0, 3}, 55], 12]
```

```
1.24993457404
```

Our trapezoidal rule error bound is

$$\text{trapezErrorBound[n\_]} = \frac{\text{bigK } (3-1)^3}{12\, n^2}$$

$$\frac{0.297507}{n^2}$$

The following table of values shows that this error bound is less than .00005 when $n \geq 78$.

```
Table[{n, trapezErrorBound[n]}, {n, 75, 80}] // TableForm
```

```
75   0.0000528901
76   0.0000515074
77   0.0000501783
78   0.0000488999
79   0.0000476697
80   0.0000464855
```

The resulting trapezoidal rule approximation is

```
NumberForm[trapez[f[x], {x, 0, 3}, 78], 12]
```

```
1.24992633699
```

For Simpson's rule we need to look at the absolute value of $f^{(4)}(x)$.

$$\text{D4f[x\_]} = \partial_{x,x,x,x}\, \text{f[x] // Simplify}$$

$$e^{-\frac{x^2}{2}} \left(3 - 6\, x^2 + x^4\right)$$

```
Plot[Abs[D4f[x]], {x, 1, 3}]
```

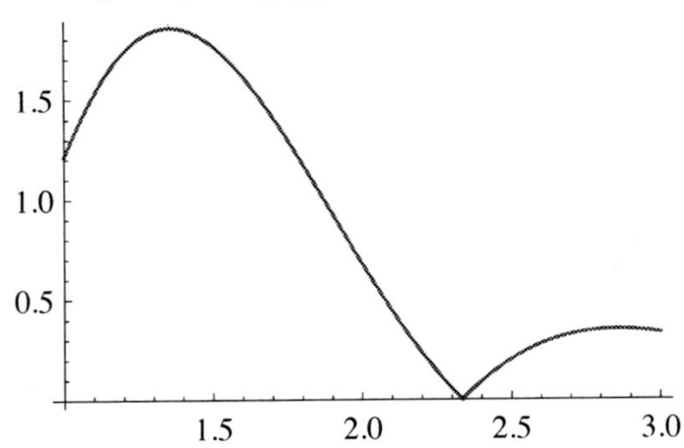

The maximum occurs at

```
FindRoot[Evaluate[∂x D4f[x]], {x, 1.4}]
```

```
{x → 1.35563}
```

and the maximum value is

```
bigM = Abs[D4f[x]] /. %
```

```
1.85487
```

The error bound for Simpson's rule with n subintervals now becomes

```
simpsonErrorBound[n_] = bigM (3 - 1)^5 / (180 n^4)
```

$$\frac{0.329755}{n^4}$$

and the following table indicates that the error bound is less than .00005 when $n \geq 10$.

```
Table[{n, simpsonErrorBound[n]}, {n, 6, 14, 2}] // TableForm
```

6	0.00025444
8	0.0000805065
10	0.0000329755
12	0.0000159025
14	8.58379×10^{-6}

The resulting Simpson's rule approximation is

```
NumberForm[simpson[f[x], {x, 0, 3}, 10], 12]
```

```
1.24992156359
```

■ NIntegrate

Mathematica has a built-in function named `NIntegrate` for numerically approximating integrals. It is used in essentially the same way as the `InputForm` of `Integrate`.

Applying `NIntegrate` to the integral $\int_0^2 \cos^2(\sin x)\, dx$ of Examples 4.4.1-4.4.4, we obtain

```
NIntegrate[Cos[Sin[x]]^2, {x, 0, 2}]
```

```
1.09868
```

■ Functions Given by Data

One of the most important uses of numerical integration is the approximation of integrals where the only information we have about the integrand is a list of discrete data points.

The **trapezoidal rule** is particularly useful when the difference between consecutive values of independent variable is not fixed. The following is a function that computes a trapezoidal rule approximation from a list of data of the form

$$\{(x_1, y_1),\ (x_2, y_2),\ \ldots,\ (x_n, y_n)\}.$$

```
trapezData[pts_] := 0.5 * Sum[(pts[[i, 2]] + pts[[i + 1, 2]]) *
    (pts[[i + 1, 1]] - pts[[i, 1]]), {i, 1, Length[pts] - 1}]
```

■ Example 4.4.6

Suppose that we have the following set of data points.

```
data = {{0, .275}, {.2, .510}, {.5, .916}, {.65, 1.186}, {1., 1.615},
   {1.3, 2.23}, {1.5, 2.703}, {1.75, 2.781}, {2., 2.607},
   {2.2, 2.562}, {2.4, 2.554}, {2.75, 2.556}, {3., 2.332},
   {3.3, 1.97}, {3.5, 1.727}, {3.7, 1.641}, {4., 1.56}, {4.25, 1.658},
   {4.6, 1.777}, {4.7, 1.686}, {5.1, 1.430}, {5.2, 1.095}, {5.4, .771},
   {5.7, 0.493}, {5.9, .383}, {6.2, .366}, {6.5, .369}, {6.75, .306},
   {7., 0.201}, {7.1, .138}, {7.5, .146}, {7.7, .166}, {8., .137},
   {8.2, 0.0737}, {8.6, .0378}, {8.75, .0529}, {9.1, .0768},
   {9.2, .0638}, {9.5, .0235}, {9.7, .00152}, {10., .0162}};
```

With `ListPlot` we can plot a *piecewise linear* approximation to the function of interest.

```
ListPlot[data, PlotJoined → True]
```

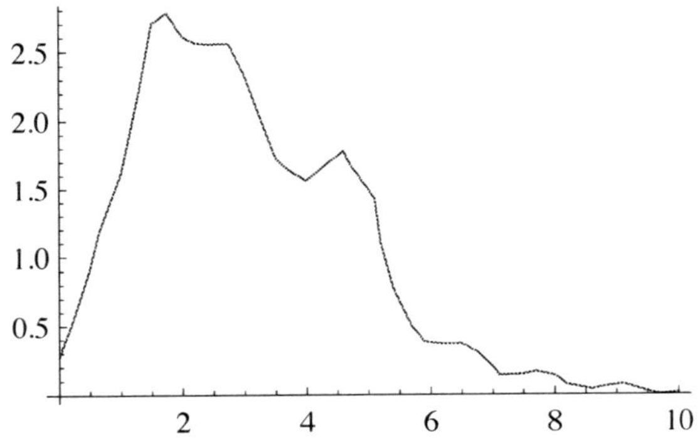

Now with our `trapezData` function we find that $\int_0^{10} f(x)\, dx$ is approximately

```
trapezData[data]
10.7437
```

We can also use Simpson's rule to approximate the integral of a function from a list of data. However, the method as it is normally derived demands that data be given at equally spaced values of the independent variable. The following function computes a Simpson's rule approximation from a list of data of the form

$$\{y_1,\ y_2,\ \ldots,\ y_n\} \text{ where } y_i = f(a + (i-1)\,h) \text{ and } n \text{ is odd.}$$

```
simpsonData[yList_, h_] := Module[{wt, n = Length[yList] - 1},
   wt[i_] := 3. + (-1)^(i-1); If[EvenQ[n],
   (h / 3) (yList[[1]] + Sum[wt[i] yList[[i]], {i, 2, n}] + yList[[n + 1]]),
   "length of data list must be odd"]]
```

■ Example 4.4.7

Suppose that we have the following list of function values at $x = 0, .25, .5, .75, \ldots, 10$.

```
data = {.0404, .0835, .0752, .162, .275, .370, .661, .916,
   1.186, 1.615, 2.033, 2.303, 2.381, 2.407, 2.562, 2.554,
   2.556, 2.332, 1.927, 1.641, 1.515, 1.358, 1.077, .886,
   .830, .795, .671, .493, .383, .366, .361, .306, .201,
   .138, .146, .1370, .0737, .0808, .0529, .0768, .0638};
```

As in the previous example, we first plot a *piecewise linear* approximation to the function of interest.

```
ListPlot[data, PlotJoined → True, Ticks → {None, Automatic}]
```

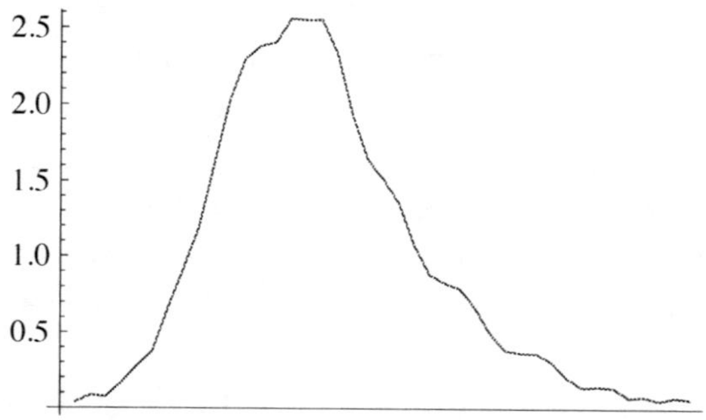

With our `simpsonData` function defined above, we find that $\int_0^{10} f(x)\,dx$ is approximately

```
simpsonData[data, .25]
```

```
9.50097
```

◆ Exercises

1. Create a `Table` with three columns and four rows that shows the midpoint, trapezoidal, and Simpson's rule approximations of the integral $\int_{-1}^{1}\sqrt{1+x^3}\,dx$ with $n = 4, 8, 16$, and 32 subintervals.

2. Let M_n, T_n, and S_n denote the approximations to $\int_a^b f(x)\,dx$, using n subintervals, obtained from the midpoint, trapezoidal, and Simpson's rules, respectively. It can be shown that

$$S_{2n} = (T_n + 2\,M_n)/3.$$

Verify this with the results of Exercise 1.

3. A factory discharges effluent into a river. The rate of effluent discharge (in cubic meters per minute) is recorded hourly over a 24-hour period. The measurements are shown in the following table.

0	1	2	3	4	5	6	7	8	9	10	11	12	13	14	15	16	17	18	19	20	21	22	23	24
3.2	2.4	4.3	5.5	5.8	6.3	4.7	4.	5.9	7.2	8.3	8.	7.3	6.6	5.1	3.8	3.5	3.1	2.2	4.1	5.5	5.1	4.8	4.1	3.7

The total effluent discharge over the 24-hour period is the integral of the rate of discharge. Convert each of the rate measurements to cubic meters per hour and use Simpson's rule to approximate the factory's total effluent discharge over this 24-hour period.

4. Modify `simpson` so that the approximation is returned in exact form unless caused to occur otherwise. (There are two occurances of "`3.`" Remove each decimal point.) Apply Simpson's rule with with $n = 4$, 8, and 16 subintervals to each of the integrals:

$$\int_0^1 x^2\,dx \qquad \int_0^1 x^3\,dx \qquad \int_0^1 x^4\,dx$$

Compare each (rational) approximation with the exact value of the integral. What do you observe? Is it surprising? Why?

4.5 Improper Integrals

The definition of the definite integral $\int_a^b f(x)\,dx$ assumes two crucial things: 1) that the interval $[a, b]$ is bounded (i.e., a and b are each finite); and 2) that the integrand f is continuous on $[a, b]$. (Actually, this second assumption can be relaxed considerably, but it is essential that f be defined and bounded on $[a, b]$.) An **improper integral** involves either an interval that is not bounded (**Type 1**) or an integrand that has a vertical asymptote at one of the endpoints of the interval (**Type 2**). In all cases, an improper integral is, by definition, a limit of definite integrals.

When f is a nonnegative function, a Type 1 improper integral can be thought of as the area of a region of infinite length, and a Type 2 improper integral can be thought of as the area of a region of infinite height.

An improper integral of Type 1 (where the interval of integration is of the form $[a, \infty)$) is defined as

$$\int_a^\infty f(x)\,dx = \lim_{t\to\infty} \int_a^t f(x)\,dx.$$

A Type 1 improper integral of the form $\int_{-\infty}^b f(x)\,dx$ is defined similarly.

An improper integral of Type 2, where f is continuous on $(a, b]$ but has a vertical asymptote at $x = a$, is defined as

$$\int_a^b f(x)\,dx = \lim_{t\to a^+} \int_t^b f(x)\,dx.$$

An improper integral of Type 2, where f is continuous on $[a, b)$ but has a vertical asymptote at $x = b$, is defined as

$$\int_a^b f(x)\,dx = \lim_{t\to b^-} \int_a^t f(x)\,dx.$$

An improper integral of any type is said to be convergent if the defining limit exists and divergent if the defining limit does not exist.

Mathematica's `Integrate` handles improper integrals automatically, but in order to get a good understanding of how improper integrals are defined, we should compute a few from scratch (more or less).

■ Example 4.5.1

A typical improper integral of Type 1 is $\int_1^\infty x^{-3/2}\,dx$. By definition,

$$\int_1^\infty x^{-3/2}\,dx = \lim_{t\to\infty}\int_1^t x^{-3/2}\,dx.$$

So first we'll compute $\int_1^t x^{-3/2}\,dx$ in terms of t and then take the limit as $t \to \infty$. The result is shown below along with a (partial) picture of the area under the graph for $x \geq 1$.

```
Assuming[t > 1, ∫_1^t x^(-3/2) dx]
Limit[%, t → ∞]
```

$2 - \frac{2}{\sqrt{t}}$

2

```
Plot[x^(-3/2), {x, 1, 15}, PlotRange → {{0, 15}, All}, Filling → Axis]
```

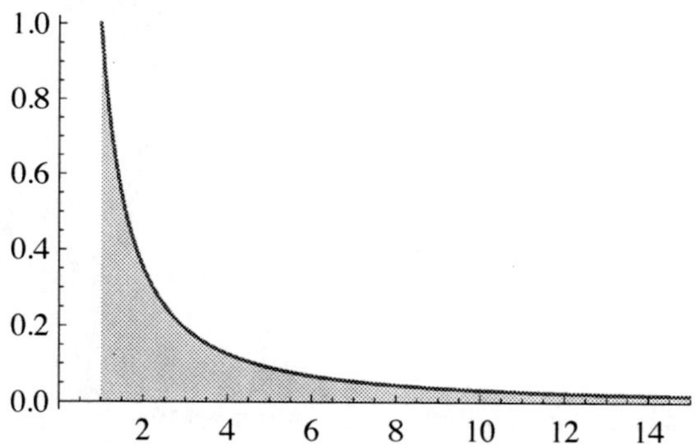

■ Example 4.5.2

A typical improper integral of Type 2 is $\displaystyle\int_0^1 \frac{1}{\sqrt{e^x - 1}}\,dx$. Note that the integrand has a vertical asymptote at $x = 0$. We'll compute the integral in terms of t and then take the limit as $t \to 0^+$.

```
Assuming[0 < t < 1, ∫_t^1 1/√(e^x - 1) dx]
Limit[%, t → 0, Direction → -1]
% // N
```

$2\left(\text{ArcTan}\left[\sqrt{-1+e}\right] - \text{ArcTan}\left[\sqrt{-1+e^t}\right]\right)$

$2\,\text{ArcTan}\left[\sqrt{-1+e}\right]$

1.83821

The following is a plot that shows the region whose area we've computed.

```
Plot[1 / √(ⅇ^x - 1), {x, 0, 1}, PlotRange → {0, 8}, Filling → Axis]
```

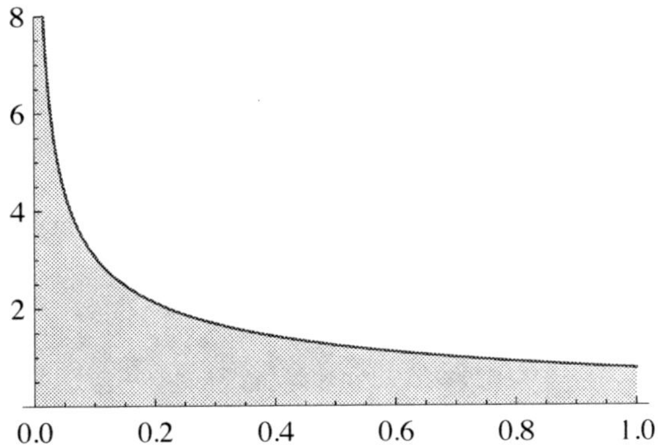

▪ Example 4.5.3

Another improper integral of Type 2 is $\int_0^{\pi/2} \tan x \, dx$. Note that the integrand has a vertical asymptote at $x = \pi/2$. We'll compute $\int_0^t \tan x \, dx$ in terms of t and then take the limit as $t \to \pi/2^+$.

```
Assuming[0 < t < π / 2, ∫₀ᵗ Tan[x] ⅆx]
Limit[%, t → π / 2]
-Log[Cos[t]]
∞
```

So this improper integral is divergent. The corresponding plot is shown below.

```
Show[Plot[Tan[x], {x, -2, 2}], Plot[Tan[x],
  {x, 0, π / 2}, Filling → Axis, PlotRange → {-12, 12}]]
```

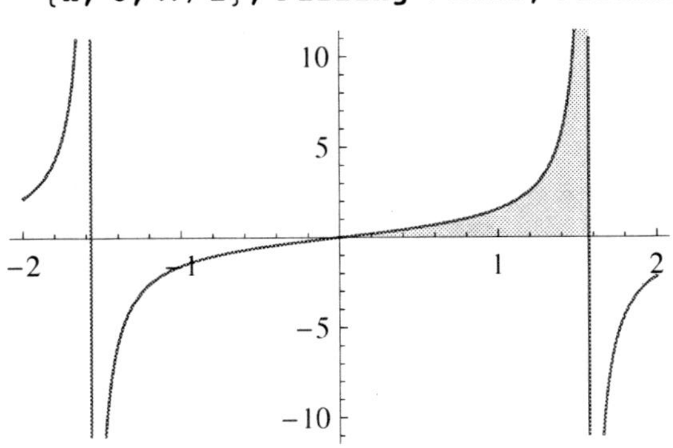

■ Example 4.5.4

Another improper integral of Type 2 is $\int_0^1 \frac{1}{\sqrt{\sqrt{x}\,(1-\sqrt{x})}}\,dx$. Note that the integrand has vertical asymptotes at both $x = 0$ and $x = 1$.

```
Plot[1 / Sqrt[Sqrt[x] (1 - Sqrt[x])], {x, 0, 1},
 PlotRange → {0, 10}, Filling → Axis]
```

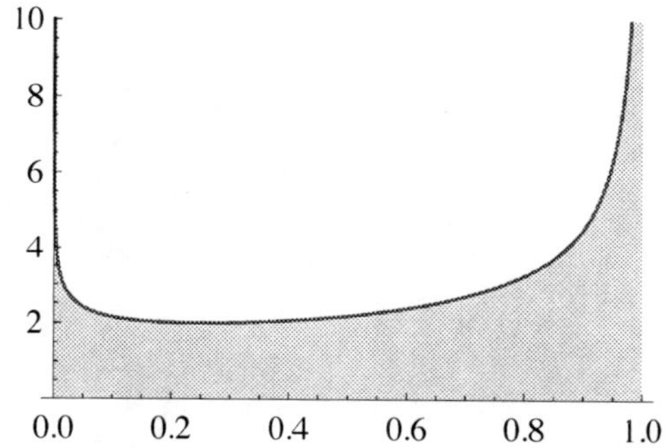

So we need to compute limits at both endpoints:

```
Assuming[0 < a < b < 1, ∫_a^b 1 / Sqrt[Sqrt[x] (1 - Sqrt[x])] dx]
Limit[Limit[%, a → 0, Direction → -1], b → 1, Direction → 1]
```

$$2\left(\sqrt{1-\sqrt{a}}\ a^{1/4} - \sqrt{1-\sqrt{b}}\ b^{1/4} - \text{ArcSin}\left[a^{1/4}\right] + \text{ArcSin}\left[b^{1/4}\right]\right)$$

π

So this peculiar region turns out to have exactly the same area as the unit disk!

As we mentioned earlier, *Mathematica*'s `Integrate` handles many improper integrals automatically. Let's recompute the improper integral in each of the preceding examples.

$$\left\{\int_1^\infty x^{-3/2}\,dx,\ \int_0^2 \frac{1}{\sqrt{e^x - 1}}\,dx,\ \int_0^{\pi/2} \text{Tan}[x]\,dx,\ \int_0^1 1\Big/\sqrt{\sqrt{x}\,\left(1-\sqrt{x}\right)}\,dx\right\}$$

Integrate::idiv: Integral of Tan[x] does not converge on $\left\{0, \frac{\pi}{2}\right\}$. ≫

$$\left\{2,\ 2\,\text{ArcTan}\left[\sqrt{-1+e^2}\right],\ \int_0^{\frac{\pi}{2}} \text{Tan}[x]\,dx,\ \pi\right\}$$

◆ Exercises

For each of the improper integrals in Exercises 1-6, find the value or determine divergence by using the appropriate limit definition. Also graph the integrand, shading the relevant region.

1. $\displaystyle\int_0^2 \frac{1}{4-x^2}\,dx$

2. $\displaystyle\int_0^2 \frac{1}{x^3+3x^2}\,dx$

3. $\displaystyle\int_0^\infty e^{-3x}\,dx$

4. $\displaystyle\int_2^\infty \frac{1}{x^3-1}\,dx$

5. $\displaystyle\int_0^\infty x\,e^{-x^2}\,dx$

6. $\displaystyle\int_0^{\pi/4} \frac{\sin x+\cos x}{\sqrt{\cos x-\sin x}}\,dx$

7. Compute each of the improper integrals

$$\int_0^1 \frac{1}{\sqrt{x}\,e^{\sqrt{x}}}\,dx \quad\text{and}\quad \int_1^\infty \frac{1}{\sqrt{x}\,e^{\sqrt{x}}}\,dx,$$

and create a shaded plot of the relevant region for each. Considering these two integrals, what should be the value of

$$\int_0^\infty \frac{1}{\sqrt{x}\,e^{\sqrt{x}}}\,dx\,?$$

8. Attempt to compute the exact value of each of

$$\int_0^1 \frac{1}{e^{\sqrt[3]{x}}-1}\,dx \quad\text{and}\quad \int_0^\infty \frac{\cos x}{1+e^x}\,dx.$$

Then use `NIntegrate` to obtain numerical values in each case.

9. A function f defined on $(-\infty,\infty)$ is a **probability density function** if:

$$\text{(i) } f(x)\ge 0 \text{ for all } x, \text{ and (ii) } \int_{-\infty}^{\infty} f(x)\,dx = 1.$$

Also, by definition,

$$\int_{-\infty}^{\infty} f(x)\,dx = \int_{-\infty}^{0} f(x)\,dx + \int_0^\infty f(x)\,dx.$$

For each of the following functions, find the number k (at least approximately) so that the function is a probability density function. Then graph the result.

a) $f(x) = \dfrac{k}{1+x^2}$

b) $f(x) = \dfrac{k}{\left(1+x^2/2\right)^{3/2}}$

c) $f(x) = k\,e^{-x^2/2}$

d) $f(x) = \begin{cases} k\,x^2\,e^{-x/2} & \text{if } x\ge 0 \\ 0 & \text{if } x\le 0 \end{cases}$

5 Applications of the Integral

Applications of integration abound in mathematics, science, engineering, economics, and numerous other fields. Here we will look at a few standard geometric problems that we hope illustrate some of underlying philosophy behind these diverse applications.

5.1 Area

The problem of finding the area under the graph of a function was our original motivation in defining the definite integral. A straightforward extension of this idea allows us to calculate the area of a region bounded by two graphs. The basic idea is that *area is the integral of the length of a typical cross-section taken perpendicular to a coordinate axis.*

■ Example 5.1.1

Find the area of the region bounded by the graphs of

```
f[x_] := x (4 - x) / 2 + Sin[5 π x / 4]; g[x_] := Sin[7 π x / 4] / 2
```

A quick plot of the two graphs reveals that they intersect at $x = 0$ and $x = 4$, which is easily confirmed by the expressions that define f and g.

```
Plot[{f[x], g[x]}, {x, -4, 8}, PlotRange → {-3, 3}]
```

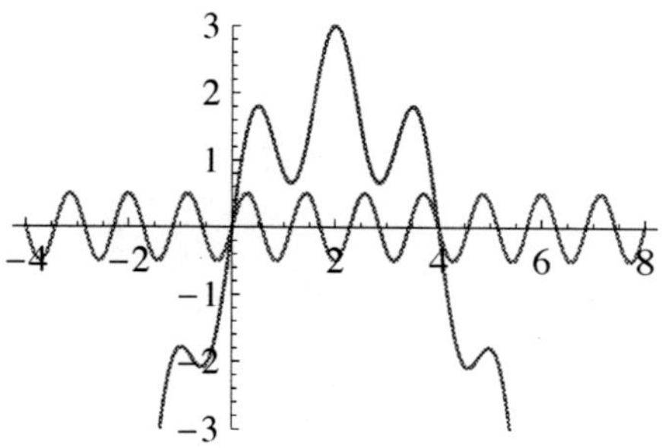

So the region of interest is the shaded region in the following plot.

```
Plot[{f[x], g[x]}, {x, 0, 4}, Filling → {1 → {2}}]
```

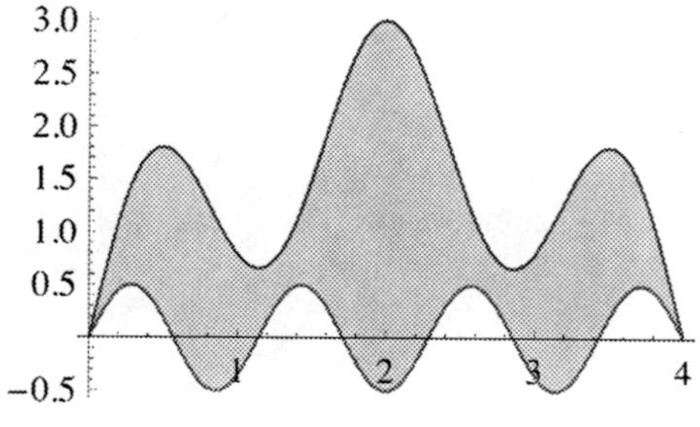

Vertical cross-sections have length given by

```
f[x] - g[x]
```

$$\frac{1}{2}(4-x)\,x+\text{Sin}\left[\frac{5\pi x}{4}\right]-\frac{1}{2}\text{Sin}\left[\frac{7\pi x}{4}\right]$$

for $0 \le x \le 4$. Thus the area of the region is

$$\int_0^4 (\mathbf{f[x] - g[x]})\, dx$$

```
% // N
```

$$\frac{16}{3}+\frac{36}{35\pi}$$

5.66074

▪ Example 5.1.2

Find the area of the region bounded by the graphs of

$$x\,y = 4 \quad \textit{and} \quad 2\,x = y\,(5 - y).$$

Let's first graph each equation, using `ContourPlot`, to get a picture of the region.

```
ContourPlot[{x y == 4, 2 x == y (5 - y)}, {x, -5, 5}, {y, -2, 6}]
```

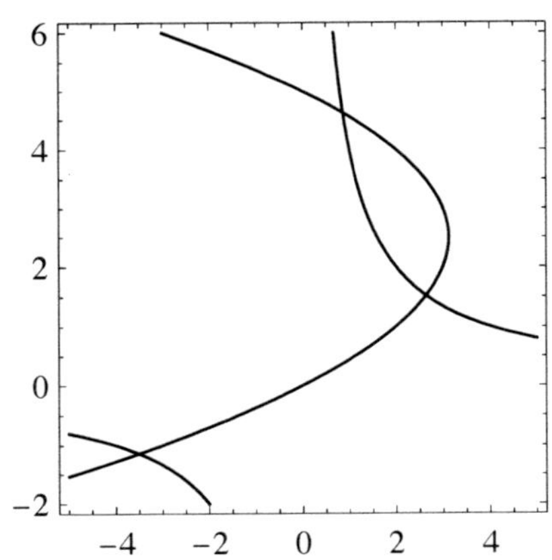

So we see that the region of interest lies in the first quadrant. Because of the particular geometry of this region, it is simpler to describe horizontal cross-sections than it is to describe vertical cross-sections. Within the region, the curve given by $2x = y\,(5 - y)$ lies to the right of the curve given by $x\,y = 4$. Thus the length of a typical horizontal cross-section is

$$\lambda[\mathtt{y\_}] := \frac{\mathbf{y\,(5 - y)}}{\mathbf{2}} - \frac{\mathbf{4}}{\mathbf{y}}$$

We will need the y-coordinates of the points of intersection, which we can find with `NSolve` as follows.

```
solns = y /. NSolve[λ[y] == 0, y]
```

{4.6262, 1.51514, -1.14134}

Clearly the positive solutions among these are the ones we are looking for. In fact, the inequality

```
solns[[2]] ≤ y ≤ solns[[1]]
```

```
1.51514 ≤ y ≤ 4.6262
```

describes the interval over which we want to integrate. So the area of the region is

$$\int_{\text{solns[[2]]}}^{\text{solns[[1]]}} \lambda[y]\, dy$$

```
3.49595
```

◆ Exercises

In Exercises 1-6, find the area of the planar region bounded by the graphs of the given equations.

1. $y = x^3$ and $y = 2x^2$
2. $y = x(2 - x)$, $y = 1 - x^2$, and $y = (x - 1)(2 - x)$
3. $y = x/2 + \sin x$ and $y = 3x^{1/4}$
4. $y = 3\sin x^2$ and $y = x^2$
5. $xy = 1$ and $x^2 + y = 3$
6. $x = y^2$ and $5y = x(x - 4)$
7. Find the area of the region inside the circle $(x - 1)^2 + y^2 = 4$ and outside the circle $x^2 + y^2 = 4$.
8. Find, in terms of a and m, the area bounded by the parabola $y = ax^2$ and the line $y = mx$.
9. Find, in terms of R and b, the area of the region inside the circle $x^2 + y^2 = R^2$ and above the line $y = b$, where $0 \le b \le R$.
10. You're stranded on a small desert island, and out of sheer boredom you decide to compute the area of your island. Beginning at one end of the island you make parallel width measurements, 50 feet apart, resulting in the following data:

    0', 127', 286', 345', 351', 344', 301', 256', 230', 258', 373', 307', 272', 243', 230', 225', 177', 0'

    Use Simpson's rule to approximate the area of the island.

5.2 Volume

In this section we will look at two examples of volume calculations for solids of revolution. The first of these examples illustrates the technique of integrating cross-sectional area to compute volume.

■ Example 5.2.1

The inside of a ten-inch tall vase can be described as the "solid" obtained by revolving the region under the graph of

```
f[x_] := 2 Sin[9 π (2 x + 5)/200]^2 + 1
```

for $0 \le x \le 10$, *about the x-axis. Find the volume of water that the vase will hold.*

```
Plot[{f[x], -f[x]}, {x, 0, 10},
 Filling → {1 → {2}}, AspectRatio → Automatic]
```

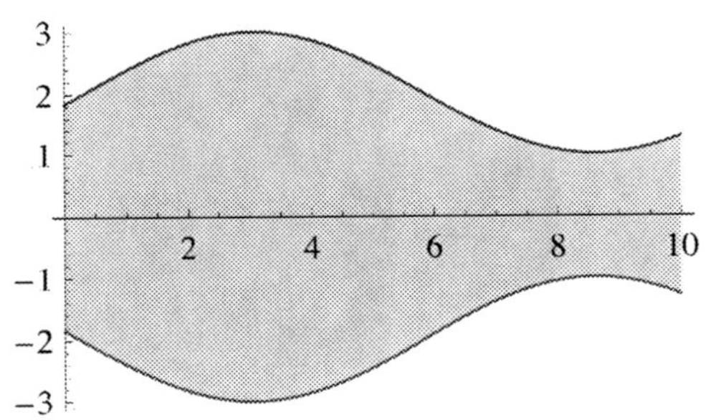

The preceding plot shows the central cross-section of the vase, parallel to the x-axis. A typical cross-section perpendicular to the x-axis will be a disk with radius $r = y$, whose area is thus given by

```
α[x] = π f[x]^2
```

$$\pi \left(1 + 2 \operatorname{Sin}\left[\frac{9}{200} \pi (5 + 2 x)\right]^2\right)^2$$

The volume inside the vase can be found by integrating this cross-sectional area from $x = 0$ to $x = 10$.

```
∫_0^10 α[x] dx // Simplify
% // N
```

$$\frac{5}{72} \left(648 \pi - 5 \left(-5 + 32 \sqrt{2} + \sqrt{5} - 64 \operatorname{Cos}\left[\frac{\pi}{20}\right]\right)\right)$$

148.567

▪ Example 5.2.2

Let $\mathcal{R}$ denote the region in the first quadrant bounded by the graphs of $y = 2\sin 2x$ and $y = x^2$. Find the volume of the solid generated by revolving $\mathcal{R}$ about the y-axis.

```
f[x_] := 2 Sin[2 x]; g[x_] := x^2
Plot[{f[x], g[x]}, {x, 0, π/2}]
```

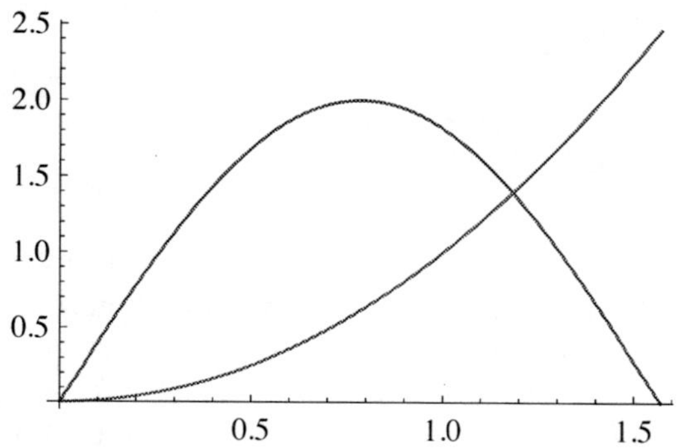

The above plot indicates the region $\mathcal{R}$. It will be important to know the intersection point of the two curves. We can find that with `FindRoot`.

```
x1 = x /. FindRoot[f[x] - g[x], {x, 1.2}]
```

```
1.18315
```

We can now create a plot that is symmetric about the *y*-axis in order to get a view of the central vertical cross-section of the solid.

```
pic = Show[Plot[{f[Abs[x]], g[x]}, {x, -x1, x1},
    PlotRange → {0, 2.3}, Filling → {1 → {2}}],
  Graphics[Circle[{0, f[x1]}, {x1, x1/5}, {π, 2 π}]],
  Graphics[Circle[{0, f[π/4]}, {π/4, π/20}]]]
```

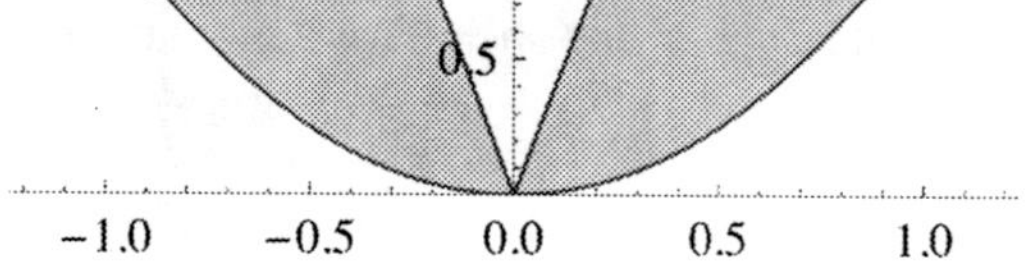

The following graphic is a more realistic rendering of the three-dimensional object.

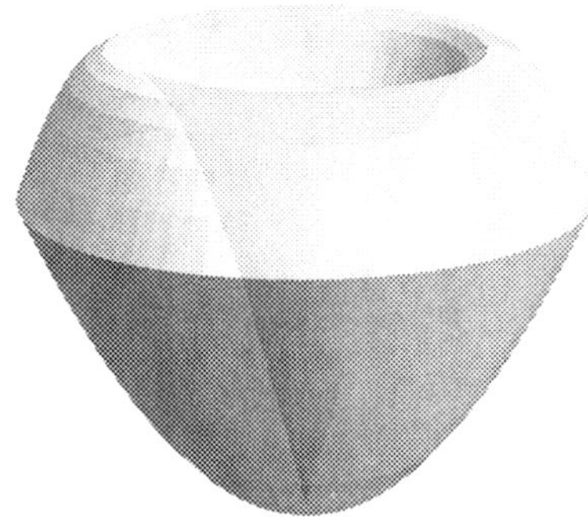

Notice that vertical cross-sections of the region $\mathcal{R}$ are much simpler to describe than horizontal cross-sections. Indeed, if we wanted to compute the area of $\mathcal{R}$, we would integrate with respect to x. So the simplest approach to calculating the volume in this example is to sum—by integration—volumes of **cylindrical shells** obtained by revolving vertical "slices" of the region $\mathcal{R}$—each with thickness dx—about the y-axis. This plot shows a typical vertical slice in $\mathcal{R}$, and beside it is a picture of the shell it generates:

```
Show[pic, Graphics[
  {Opacity[.3], Rectangle[{.55, g[.6]}, {.65, f[.6]}]}]]
```

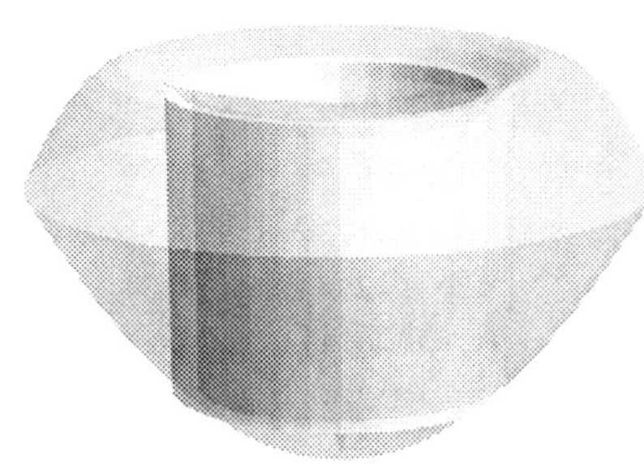

Since the height of the slice—and therefore the shell—is $f(x) - g(x)$, the volume of a typical cylindrical shell with radius x is

$$dV = 2\pi x\,(f(x) - g(x))\,dx.$$

Note that the $2\pi x$ factor represents the circumference of the shell. Now, by integrating this over $[0, x_1]$, we obtain the volume of the solid:

$$\int_0^{x1} 2\,\pi\,x\,(f[x] - g[x])\,dx$$

```
4.43025
```

◆ Exercises

In Exercises 1-4, find the volume of the solid generated by revolving about the x-axis the region bounded by the given graphs.

1. $y = x^3$ and $y = 2x^2$
2. $y = x(2 - x)$, $y = 1 - x^2$, and $y = (x - 1)(2 - x)$
3. $y = x/2 + \sin x$, $x = 2\pi$, and $y = 0$
4. $xy = 1$ and $x^2 + y = 3$

In Exercises 5-8, find the volume of the solid generated by revolving about the y-axis the region bounded by the given graphs.

5. $y = x^3$ and $y = 2x^2$
6. $y = x(2 - x)$, $y = 1 - x^2$, and $y = (x - 1)(2 - x)$
7. $x = y^2$ and $5y = x(x - 4)$
8. $xy = 1$ and $x^2 + y = 3$
9. Find, in terms of R and h, the volume of the solid generated by revolving the region inside the circle $x^2 + y^2 = R^2$, to the right of the y–axis and below the line $y = h - R$, where $0 \leq h \leq R$.

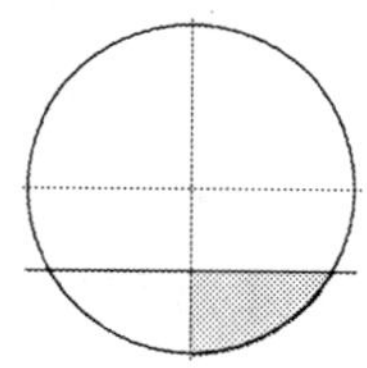

10. Find the volume of the *torus* generated by revolving each of the following disks about the y–axis.
    a) $(x - 1)^2 + y^2 \leq 1$
    b) $(x - 3)^2 + y^2 \leq 4$
    c) $(x - R)^2 + y^2 \leq r^2$ where $0 \leq r \leq R^2$

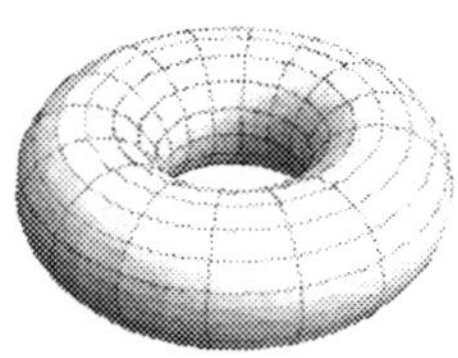

5.3 Arc Length and Surface Area

■ Arc Length

The length of a smooth curve described by $y = f(x)$, $a \leq x \leq b$, is given by the integral

$$\int_a^b \sqrt{1 + (f'(x))^2}\, dx.$$

Mathematica is especially useful for this type of problem, since antiderivatives for the integrands in integrals such as this are typically very difficult and often impossible to find in terms of elementary functions, even when the function f is quite simple.

Example 5.3.1

Find the length of the curve $y = \sin x,\ \ 0 \le x \le \pi$.

```
Plot[Sin[x], {x, 0, π}, AspectRatio → Automatic]
```

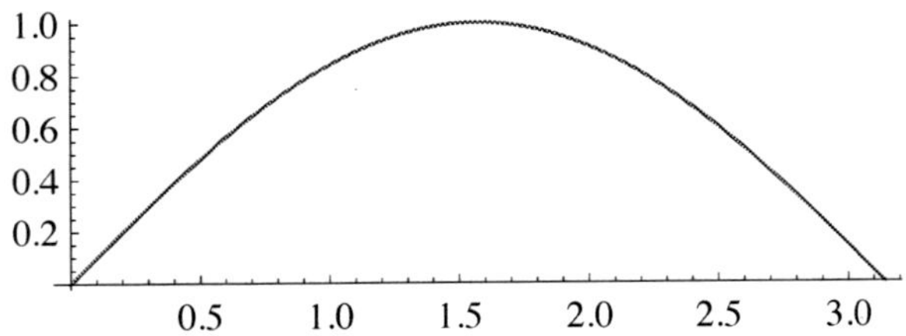

The derivative here is $f'(x) = \cos x$, and so the arc length between 0 and b is given by the special function `EllipticE`.

```
arclen[b_] = Assuming[b > 0, ∫_0^b √(1 + Cos[x]^2) ⅆx]
```

$\sqrt{2}$ `EllipticE[b, 1/2]`

The desired arclength is the value at $b = \pi$.

```
arclen[π] // N
```

```
3.8202
```

■ Example 5.3.2

Find the length of the curve $y = x^3,\ \ 0 \le x \le 1$.

```
Plot[x^3, {x, 0, 1}, AspectRatio → Automatic]
```

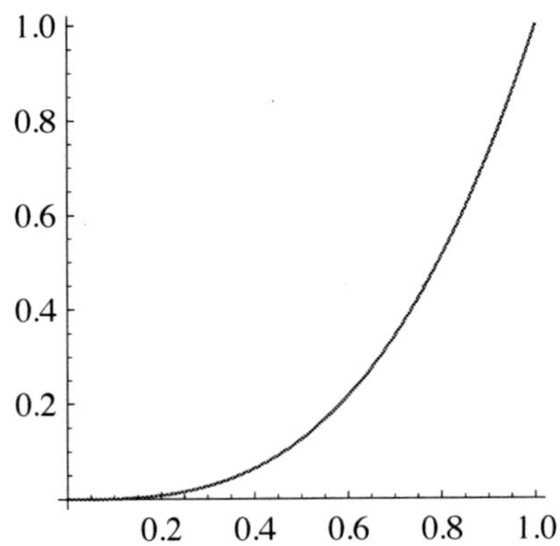

The derivative here is $f'(x) = 3x^2$, and so the arc length is given by the integral $\int_0^1 \sqrt{1+9x^4}\ dx$, which will, at best, be given by *Mathematica*'s `Integrate` in terms of some special function (as in Example 5.3.1). So we may as well use `NIntegrate` instead:

```
NIntegrate[√(1 + 9 x^4), {x, 0, 1}]
```

```
1.54787
```

■ Surface Area

The areas of the surfaces generated by revolving a curve $y = f(x)$, $a \le x \le b$, about the x- and y-axes, respectively, are

$$\int_a^b 2\pi f(x)\sqrt{1+(f'(x))^2}\,dx \quad\text{and}\quad \int_a^b 2\pi x\sqrt{1+(f'(x))^2}\,dx.$$

Likewise, the areas of the surfaces generated by revolving a curve $x = f(y)$, $c \le y \le d$, about the x- and y-axes, respectively, are

$$\int_c^d 2\pi y\sqrt{1+(f'(y))^2}\,dy \quad\text{and}\quad \int_c^d 2\pi f(y)\sqrt{1+(f'(y))^2}\,dy.$$

▪ Example 5.3.3

Find the surface area generated by revolving the arc $y = \sin x$, $0 \le x \le \pi$, about the x-axis.

$$\int_0^\pi 2\,\pi\,\mathbf{Sin}[\mathbf{x}]\,\sqrt{1+\mathbf{Cos}[\mathbf{x}]^2}\,d\mathbf{x}$$

```
% // N
```

$$2\,\pi\left(\sqrt{2}+\text{ArcSinh}[1]\right)$$

```
14.4236
```

▪ Example 5.3.4

Find the surface area generated by revolving the arc $y = \sin x$, $0 \le x \le \pi$, about the y-axis.

$$\mathbf{NIntegrate}\left[2\,\pi\,\mathbf{x}\,\sqrt{1+\mathbf{Cos}[\mathbf{x}]^2},\ \{\mathbf{x},\,0,\,\pi\}\right]$$

```
37.7038
```

◆ Exercises

In Exercises 1–4, find:

a) the length of the given arc;

b) the area of the surface generated by revolving the arc about the x-axis;

c) the area of the surface generated by revolving the arc about the y-axis.

1. $y = 1/x$, $1 \le x \le 2$
2. $y = \frac{1}{2}(e^x + e^{-x})$, $0 \le x \le 1$
3. $y = \tan x$, $1 \le x \le \pi/3$
4. $y = \sqrt{x}$, $0 \le x \le 1$
5. Find, in terms of R and h, the area of the surface generated by revolving about the y-axis the portion of the circle $x^2 + y^2 = R^2$, to the right of the y-axis and below the line $y = h - R$, where $0 \le h \le 2R$.

6. Using Simpson's rule and the following table of values, approximate

   a) the length of the arc $y = f(x), 0 \le x \le 5$;

   b) the area of the surface generated by revolving the arc about the x-axis;

   c) the area of the surface generated by revolving the arc about the y-axis.

x	0	0.5	1	1.5	2	2.5	3	3.5	4	4.5
$f(x)$	1.2	3.1	3.9	4.5	5.3	4.7	4.3	3.3	2.5	1.3
$f'(x)$	3.5	2.3	1.6	1.3	−0.2	−1.1	−1.5	−2.1	−1.9	−1.2

5.4 Moments and Centroids

Consider a region $\mathcal{R}$ in the plane, with area A, bounded by $y = f(x)$, $y = g(x)$, $x = a$, and $x = b$, where $f(x) > g(x)$ for $a \le x \le b$. If a plate with uniform thickness and constant mass density occupies $\mathcal{R}$, then the plate's **center of mass** coincides with the **centroid** of $\mathcal{R}$, which is the point

$$(\overline{x}, \overline{y}) = \left(M_y / A, M_x / A\right),$$

where M_x and M_y are its **moments** about the x- and y-axes given by

$$M_x = \frac{1}{2}\int_a^b \left[f(x)^2 - g(x)^2\right] dx \quad \text{and} \quad M_y = \int_a^b x\,(f(x) - g(x))\, dx$$

▪ Example 5.4.1

Find the centroid of the region in the first quadrant bounded by $y = 4\,e^{x-2}$ *and* $y = x^2$.

```
f[x_] := 4 e^(x-2); g[x_] := x^2
region = Plot[{f[x], g[x]}, {x, 0, 2}, Filling → {1 → {2}}]
```

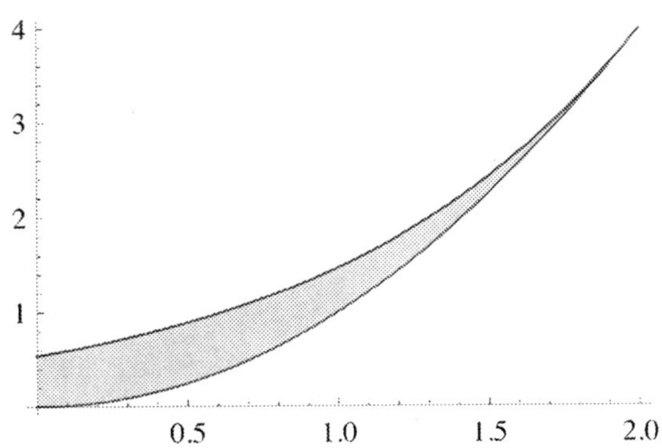

It is easy to observe that the point of intersection of the two graphs is (1, 1). The area of the region is

```
area = ∫₀² (f[x] - g[x]) ⅆx
```

$$\frac{4}{3}-\frac{4}{e^2}$$

Next we'll calculate the moments M_y and M_x:

```
xMoment = 1/2 ∫₀² (f[x]² - g[x]²) ⅆx
```

$$\frac{1}{2}\left(\frac{8}{5}-\frac{8}{e^4}\right)$$

```
yMoment = ∫₀² x (f[x] - g[x]) ⅆx
```

$$\frac{4}{e^2}$$

Finally, we complete the calculation of $(\overline{x}, \overline{y})$ and show its location.

```
{yMoment/area, xMoment/area} // Simplify
centroid = % // N
```

$$\left\{\frac{3}{-3+e^2}, \frac{3\left(-5+e^4\right)}{5 e^2\left(-3+e^2\right)}\right\}$$

```
{0.683518, 0.917607}
```

```
Show[region, Graphics[{PointSize[.02], Point[centroid]}]]
```

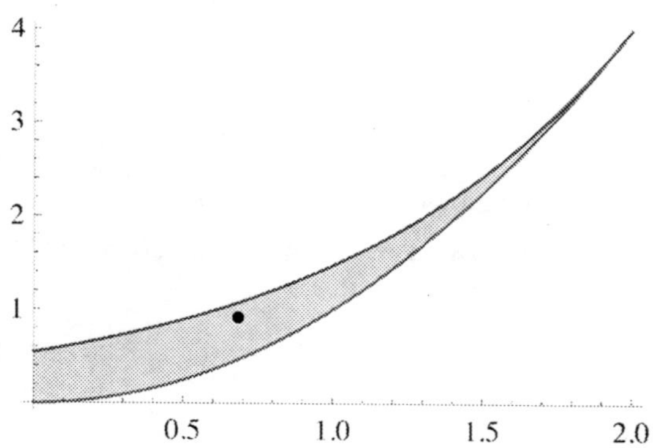

■ Example 5.4.2

A boomerang with uniform thickness and constant mass has the shape of the region bounded by the graphs of $y = \sin x$ and $y = 0.9 \sin^4 x$ between $x = 0$ and $x = \pi$. Find its center of mass.

```
f[x_] := Sin[x]; g[x_] := 9 / 10 Sin[x]^4
region = Plot[{f[x], g[x]}, {x, 0, π}, Filling → {1 → {2}}]
```

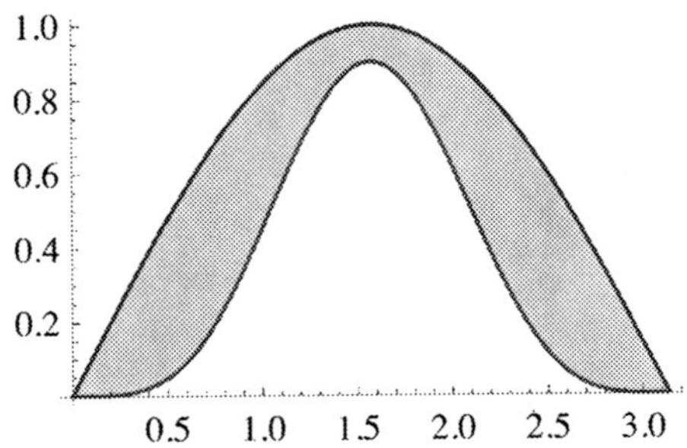

It is obvious that $\overline{x} = \pi/2$ because of symmetry. So we only need to find $\overline{y}$. The calculation proceeds as follows.

$$\textbf{area} = \int_0^{\pi} (\textbf{f[x]} - \textbf{g[x]})\, dx$$

$$2 - \frac{27\pi}{80}$$

$$\textbf{xMoment} = \frac{1}{2}\int_0^{\pi} \left(\textbf{f[x]}^2 - \textbf{g[x]}^2\right) dx$$

$$\frac{713\pi}{5120}$$

$$\left\{\frac{\pi}{2}, \frac{\textbf{xMoment}}{\textbf{area}}\right\} \text{ // Simplify}$$

```
ctrOfMass = % // N
```

$$\left\{\frac{\pi}{2}, \frac{713\pi}{10240 - 1728\pi}\right\}$$

```
{1.5708, 0.465559}
```

Finally, we replot the region, showing also the location of the centroid.

```
Show[region, Graphics[{PointSize[Medium], Point[ctrOfMass]}]]
```

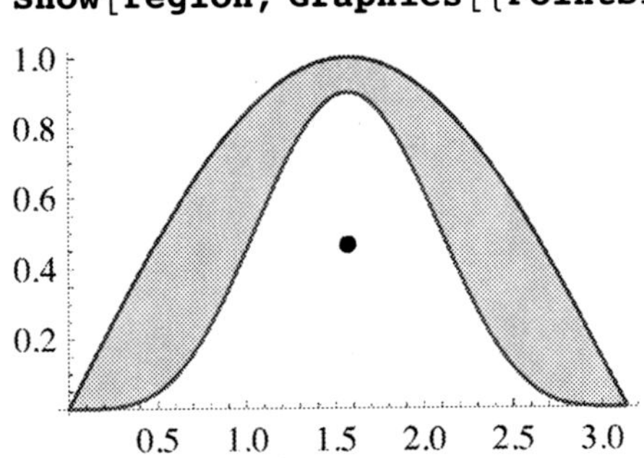

◆ Exercises

In Exercises 1-5, find the centroid of the region bounded by the graphs of the given equations. Use symmetry whenever possible. Show the location of the centroid in a shaded plot each region.

1. $y = x^2,\ y = 4$
2. $y = x^2,\ y = x$
3. $y = e^{x-1},\ y = x^2$
4. $y = \cos x,\ y = 0,\ x = \pm\frac{\pi}{2}$
5. $y = \sin x,\ y = \cos x,\ x = 0,\ x = \frac{\pi}{4}$

In Exercises 6-10, find the centroid of the described region. Use symmetry whenever possible.

6. The quarter of the unit disk $x^2 + y^2 \le 1$ in the first quadrant
7. The top half of the unit disk $x^2 + y^2 \le 1$
8. The triangle in the first quadrant under the line $x + y = 1$
9. The square $-1 \le x \le 1,\ 0 \le y \le 2$, surmounted by a half-disk
10. The portion of the unit disk above the line $y = 1/2$
11. Think of the region in Example 1 as a flat plate with uniform thickness and constant mass density. Imagine also that the x-axis is a flat surface upon which the plate is balanced on its edge. Clearly the plate must be supported in some way in order to be positioned as it is, and if the plate were no longer supported, it would roll along its bottom edge toward the right and come to rest in some "equilibrium position." The question is this: In what position will the plate come to rest if it is no longer supported? In particular, what will be the new coordinates of its center of mass?
12. Create a function `centroid[f_,g_,{x_,a_,b_}]`, that returns the centroid of the region bounded by $y = f(x)$ and $y = g(x)$ between $x = a$ and $x = b$, assuming that $f(x) \ge g(x)$ for $a \le x \le b$. Do the appropriate calculations inside a `Module` and use `NIntegrate` to do the integration. Test the function on Examples 5.4.1 and 5.4.2. Then use it to rework Exercises 1-5 above.

6 Differential Equations

Many real-world phenomena can be described in terms of relationships between quantities and their rates of change, and such relationships give rise to differential equations. In this section we will explore some of the ways that *Mathematica* can be used to study differential equations.

6.1 Equations and Solutions

The differential equations we will consider here can be written in the form

$$\frac{dy}{dt} = f(t, y),$$

where t is the *independent variable*, y is the *dependent variable*, and f is a given continuous function of two variables. Such an equation is called a *first-order* differential equation. An equation such as this is said to be a first-order *linear* differential equation if it can be written in the form

$$\frac{dy}{dt} + p(t)\, y = q(t),$$

where p and q are given, possibly nonlinear, functions of one variable.

A function $y(t)$ is a **solution of** $\frac{dy}{dt} = f(t, y)$ **on an interval** I if $y'(t) = f(t, y(t))$ for all t in I.

■ Example 6.1.1

Verify that $y(t) = \left(1 - t^2\right)^{-1}$ *is a solution of* $\frac{dy}{dt} = 2\, t\, y^2$ *on any interval that does not contain* $t = \pm\, 1$. *Also plot* $y(t)$ *and describe the largest intervals on which* $y(t)$ *is a solution.*

After entering the function

```
y[t_] := 1/(1 - t^2)
```

we can verify the solution either by computing and comparing each side of the differential equation,

```
{y'[t], 2 t y[t]^2}
```

$$\left\{\frac{2\,t}{\left(1-t^2\right)^2}, \frac{2\,t}{\left(1-t^2\right)^2}\right\}$$

or by computing the difference $y'(t) - 2\, t\, y(t)$,

```
y'[t] - 2 t y[t]^2
0
```

or by checking the truth of the differential equation:

```
y'[t] == 2 t y[t]^2
True
```

Here is the graph of $y(t)$:

```
Plot[y[t], {t, -3, 3}, PlotRange → {-8, 8}]
```

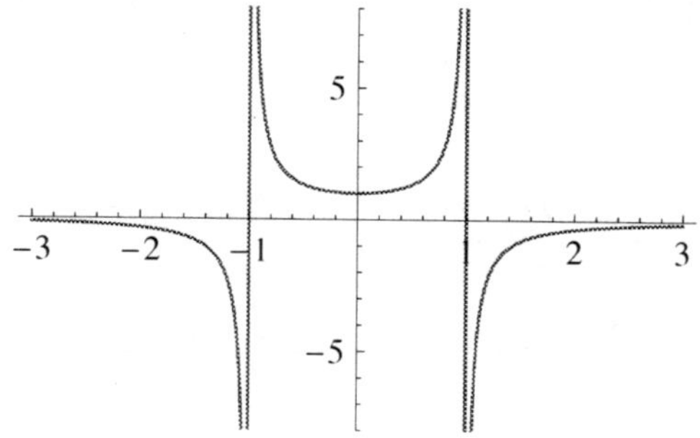

Since $y(t)$ is undefined at $t = \pm 1$, we must conclude that the equation is satisfied only when $t \neq \pm 1$. Therefore, $y(t)$ is a solution of the differential equation on an interval I only if I does not contain ± 1. The largest such intervals are $(-\infty, -1)$, $(-1, 1)$, and $(1, \infty)$. The plot of $y(t)$ on $(-\infty, \infty)$ does *not* represent a solution of the differential equation on $(-\infty, \infty)$. It shows three solutions of the differential equation on three different intervals.

■ Example 6.1.2

Verify that $y(t) = 2\left(t^2+1\right)^{-2} + t^2 + 1$ *is a solution of the linear differential equation*

$$\frac{dy}{dt} + \frac{4t}{t^2+1}\,y = 6t$$

on any interval. Also plot the solution.

We'll first enter the function and check that it satisfies the differential equation for all t.

```
y[t_] := 2 (t^2 + 1)^-2 + t^2 + 1

y'[t] + (4 t)/(t^2 + 1) y[t] // Simplify

6 t
```

The following creates the desired plot.

```
Plot[y[t], {t, -2, 2}, PlotRange → {0, 5}]
```

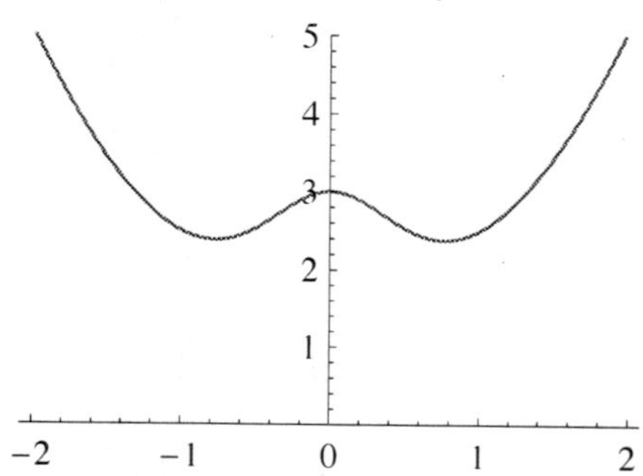

It is clear in this case that $y(t)$ satisfies the differential equation for all t and thus is a solution on any interval.

Differential equations typically have many solutions. First-order equations typically have a *one-parameter family* of solutions. This one-parameter family is called the **general solution** of the differential equation. In each of the following examples, the constant C represents the parameter.

■ Example 6.1.3

Verify that $y(t) = (C - t^2)^{-1}$ *is a solution of* $\frac{dy}{dt} = 2\,t\,y^2$ *on any interval in which* $C - t^2 \neq 0$. *Plot solutions corresponding to* $C = -.25, -.1, 0, .25, 1, 4$.

We'll first enter the function and check that it satisfies the differential equation everywhere except at $t = \pm C$.

```
y[t_] := 1/(c - t^2)
```

```
y'[t] - 2 t y[t]^2
```

```
0
```

The following creates the desired plot.

```
solns = y[t] /. c → {-.25, -.1, 0, .5, 2, 4}
```

$$\left\{\frac{1}{-0.25-t^2}, \frac{1}{-0.1-t^2}, -\frac{1}{t^2}, \frac{1}{0.5-t^2}, \frac{1}{2-t^2}, \frac{1}{4-t^2}\right\}$$

```
Plot[Evaluate[solns], {t, -2.5, 2.5}, PlotRange → {-11, 10}]
```

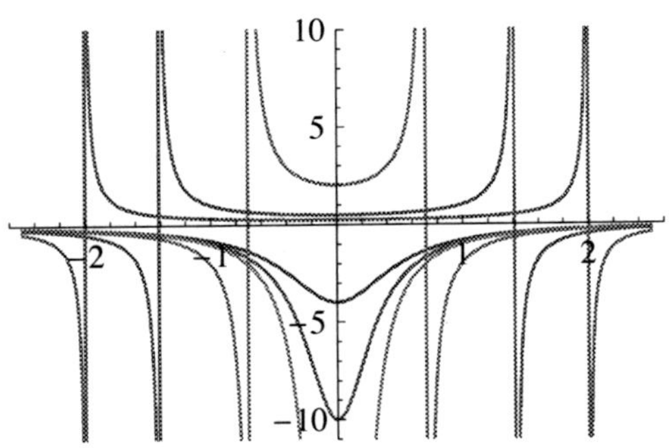

■ Example 6.1.4

Verify that $y(t) = C\,(t^2+1)^{-2} + t^2 + 1$ *is a solution of*

$$\frac{dy}{dt} + \frac{4t}{t^2+1}\,y = 6t$$

on any interval. Plot solutions corresponding to $C = -1, 0, 1, 2, 3, 4$.

First we enter the function and check that it satisfies the differential equation for all t, and consequently on any interval.

```
y[t_] := c (t^2 + 1)^-2 + t^2 + 1
```

```
y'[t] + 4 t/(t^2 + 1) y[t] // Simplify
```

```
6 t
```

The desired plot is created as follows.

```
y[t] /. c → {-4, -2, 0, 2, 4}
Plot[Evaluate[%], {t, -2, 2}]
```

$$\left\{1+t^2-\frac{4}{(1+t^2)^2},\ 1+t^2-\frac{2}{(1+t^2)^2},\ 1+t^2,\ 1+t^2+\frac{2}{(1+t^2)^2},\ 1+t^2+\frac{4}{(1+t^2)^2}\right\}$$

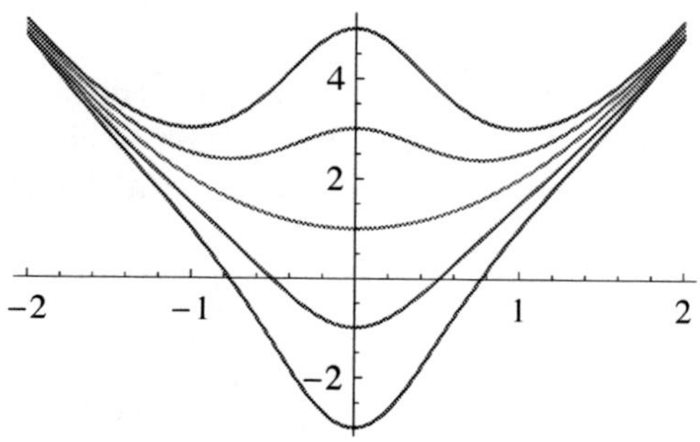

Initial Value Problems

A first-order **initial value problem** consists of a first-order differential equation and an *initial value*:

$$\frac{dy}{dt} = f(t, y), \quad y(t_0) = y_0.$$

Note that the initial value simply requires that the graph of the solution pass through the point (t_0, y_0).

Example 6.1.5

Verify that $y = 4\left(1 - 4t^2\right)^{-1}$ *is a solution of*

$$\frac{dy}{dt} = 2\,t\,y^2, \quad y(0) = 4.$$

on the interval $-.5 < t < .5$. *Also plot the solution.*

Here we'll do essentially the same thing as in Example 6.1.1 to verify that y satisfies the differential equation (for all $t \neq .5$). Then we'll do a simple evaluation to check the initial condition.

```
y[t_] := 4 / (1 - 4 t^2)
y'[t] - 2 t y[t]^2
```

```
0
```

```
y[0]
```

```
4
```

```
Plot[y[t], {t, -.5, .5}, PlotRange → {0, 20}]
```

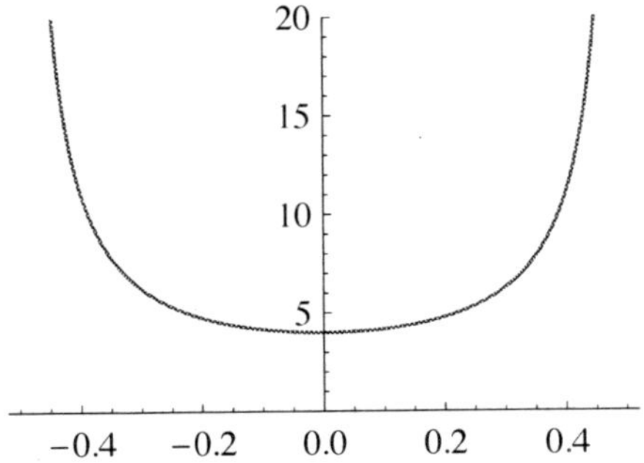

Note that y satisfies the differential equation at all $t \neq \pm .5$; so it cannot be a solution of the differential equation on an interval that contains $t = \pm .5$. Therefore, since the initial value is given at $t = 0$, y cannot satisfy the initial value problem on any interval larger than $-.5 < t < .5$.

■ Example 6.1.6

Verify that $y = t^2 + 1 + 8\left(t^2 + 1\right)^{-2}$ *is a solution of*

$$\frac{dy}{dt} + \frac{4t}{t^2 + 1}\, y = 6t, \quad y(1) = 4,$$

for all t. *Also plot the solution.*

The following computation shows that the differential equation is satisfied for all t.

```
y[t_] := t^2 + 1 + 8 (t^2 + 1)^-2
y'[t] + (4 t)/(t^2 + 1) y[t] // Simplify
6 t
```

Also, the initial condition is satisfied:

```
y[1]
4
```

The graph of the solution is as follows:

```
Plot[y[t], {t, -4, 4}, PlotRange → {0, 15}]
```

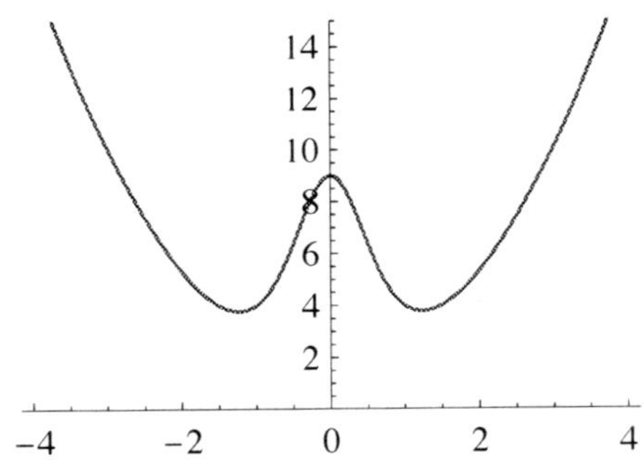

Finding the solution of an initial value problem is a matter of selecting the one member of the general solution family that satisfies the given initial condition. Such a solution can often be obtained by substituting the given initial values of t and y into the general solution and then solving for the parameter C. The next example illustrates this.

■ Example 6.1.7

Verify that $y = 10/\left(1 + C\, e^{-t/2}\right)$ *is a solution of*

$$\frac{dy}{dt} = \frac{y}{20}(10 - y)$$

for any constant C. Then find and plot the solution of the initial value problem

$$\frac{dy}{dt} = \frac{y}{20}(10 - y), \quad y(0) = 2\,.$$

First we'll verify that y satisfies the differential equation for any C (for all t where $1 + C\,e^{-t/2} \neq 0$):

```
y[t_] := 10/(1 + c e^(-t/2))

y'[t] - y[t]/20 (10 - y[t]) // Simplify
```

0

Now we set $y(0) = 2$, and solve for C.

```
Solve[y[0] == 2, c] // Flatten
```

{c → 4}

So the solution is

```
y[t] = y[t] /. %
```

$$\frac{10}{1 + 4\, e^{-t/2}}$$

and its graph is as follows:

```
Plot[y[t], {t, -10, 15}]
```

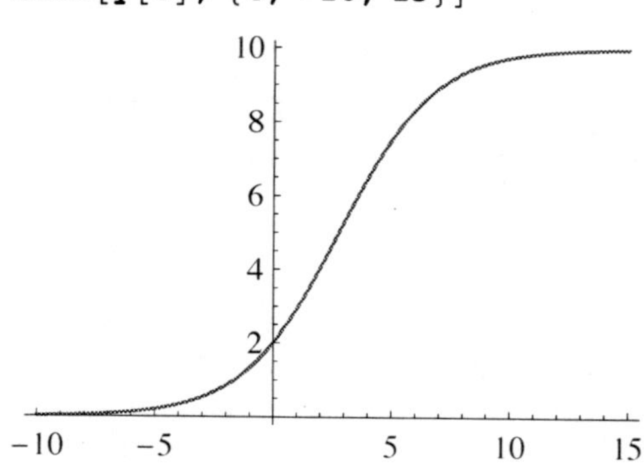

(Notice the horizontal asymptotes at $y = 0$ and $y = 10$, which are precisely the zeros of the right side of the differential equation. Why should this be so?)

◆ Exercises

In each of Exercises 1-5, check that the family of functions described in the first column satisfies the differential equation in the second column. Then plot the solution for each value of C in the third column on the interval given in the fourth column.

1. $y = C e^{-3t}$	$\frac{dy}{dt} + 3y = 0$	$C = \pm 1, \pm 2$	$[-.25, 1]$
2. $y = \frac{2}{1 + C e^{-2t}}$	$\frac{dy}{dt} - 2y = -y^2$	$C = -9, -.5, 0, 3$	$[0, 2]$
3. $y = \frac{(1+3t)^{4/3} + C}{(1+3t)^{1/3}}$	$\frac{dy}{dt} + \frac{y}{1+3t} = 4$	$C = 0, 25, 50$	$[0, 10]$
4. $y = t^2(C + \ln t)$	$\frac{dy}{dt} - 2y/t = t$	$C = 1, -2, -3$	$[0, 25]$
5. $y = \sin^{-1}(C e^{-t})$	$\frac{dy}{dt} + \tan y = 0$	$C = \pm 1, \pm \frac{1}{2}$	$[-1, 2]$

In Exercises 6-10, use the general solution from the corresponding Exercise 1-4 to solve the given initial value problem. Plot the solution.

6. $\frac{dy}{dt} + 3y = 0, \quad y(0) = 5$

7. $\frac{dy}{dt} + 3y = 0, \quad y(0) = 5$

8. $\frac{dy}{dt} - 2y = -y^2, \quad y(0) = 5$

9. $\frac{dy}{dt} - 2y/t = t, \quad y(1) = -\frac{1}{2}$

6.2 Direction Fields

If the graph of a solution of the differential equation

$$\frac{dy}{dt} = f(t, y)$$

passes through a point (t_1, y_1), then the slope of the graph at that point is $m = f(t_1, y_1)$. Thus, we can think of the function f as specifying a *slope field*, or *direction field*, in the ty-plane. This direction field can be visualized as an array of arrows, each located at a point (t, y) with slope given by $f(t, y)$. The plot below shows the direction field given by $f(t, y) = t - y^2$ together with a few solution curves.

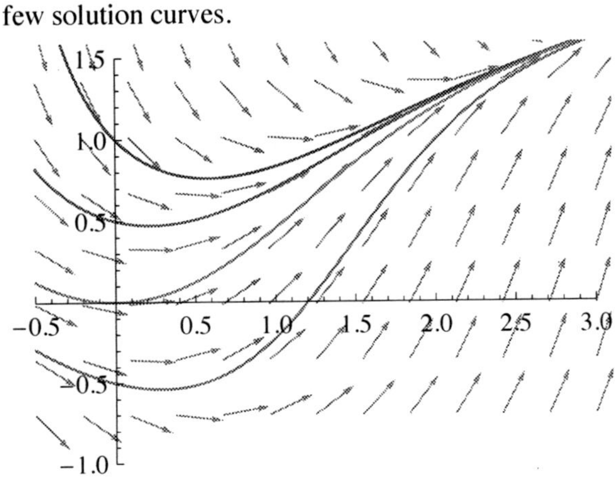

The tool that we need to plot direction fields is in *Mathematica*'s `VectorFieldPlots` package.

```
Needs["VectorFieldPlots`"]
```

This package provides a function called `VectorFieldPlot` that can be used to plot the direction field of a differential equation. The direction field shown above (for $f(t, y) = t - y^2$) can be plotted, with *no options*, as follows. Notice that the first argument takes the form `{1, ` $f(t, y)$`}`.

```
VectorFieldPlot[{1, t - y²}, {t, -.5, 3}, {y, -.7, 1.7}]
```

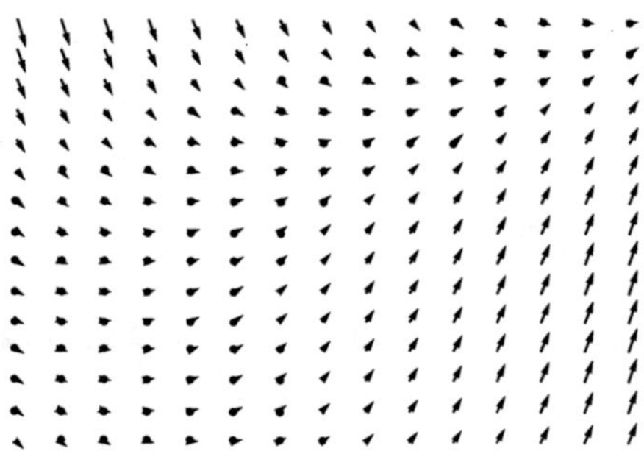

A much better plot results from supplying various options. In particular, we would like for all of the arrows to have the same length and for the axes to be shown. The following defines a function named `plotDirectionField` that incorporates these options, as well as others that produce red arrows and fewer arrows in each direction. These default options can be overridden, and additional options can be given. (Notice the fourth argument, `options`, which is followed by three `Blanks`.) Also notice that the first argument is simply an expression for $f(t, y)$.

```
plotDirectionField[fn_, {x_, a_, b_}, {y_, c_, d_}, options___] :=
 VectorFieldPlot[{1, Evaluate[fn /. x → t /. y → w]},
  {t, a, b}, {w, c, d}, options, ScaleFunction → (1 &),
  PlotPoints → {13, 8}, ColorFunction → (Darker[Gray] &)]
```

This is the direction field in the first plot shown above:

```
plotDirectionField[t - y², {t, -.5, 3}, {y, -.7, 1.7}]
```

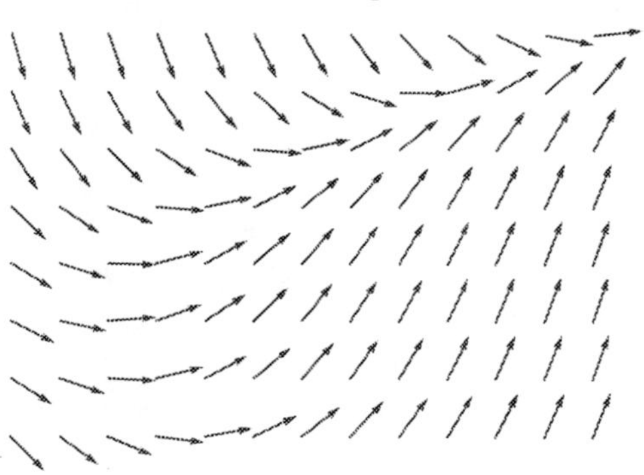

■ Example 6.2.1

Verify that the general solution of

$$\frac{dy}{dt} = -t\,y$$

is $y(t) = C\,e^{-t^2/2}$. *Then plot the direction field for the differential equation along with solution curves corresponding to* $C = -2.5, -1.5, \ldots, 2.5$.

First we'll define $y(t)$ and check that it satisfies the differential equation for any C.

```
y[t_] := c e^(-t^2/2)

y'[t] + t y[t]
```

```
0
```

The direction field is created by entering

```
dirfield = plotDirectionField[-t y, {t, -3, 3}, {y, -2.25, 2.25}]
```

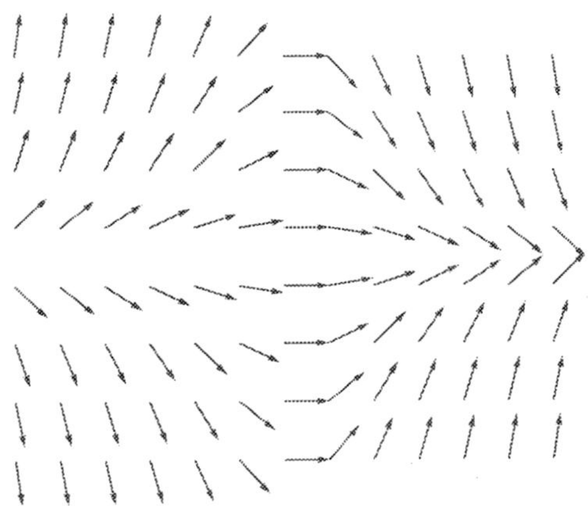

and the requested collection of curves is plotted by entering

```
curves = Plot[Evaluate[Table[c e^(-t^2/2), {c, -2.5, 2.5, 1}]],
  {t, -3, 3}, PlotStyle -> Thickness[.004]]
```

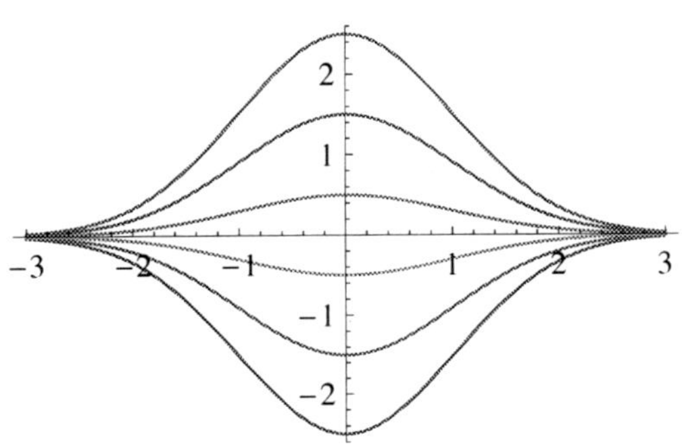

Finally, we show the curves on top of the direction field:

```
Show[dirfield, curves, Axes → True]
```

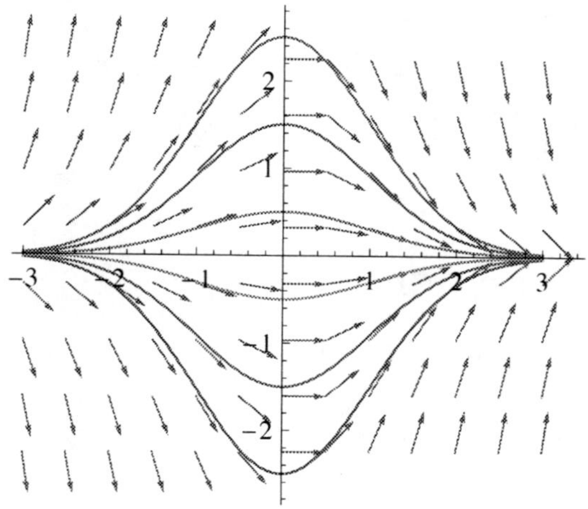

Exercises

For each of Exercises 1-5 in Section 6.1, replot all of the requested curves on top of the direction field for the differential equation.

6.3 Euler's Method

There are numerous techniques for finding exact solutions of differential equations. Different types of equations require different techniques, two of which we will see in the next section. But situations in which exact solutions are extremely difficult or impossible to find are quite common. So numerical methods for approximating solutions are crucial.

A numerical method for approximating the solution of an initial value problem

$$\frac{dy}{dt} = f(t, y), \quad y(t_0) = y_0$$

typically consists of a recurrence formula for computing an approximation at time $t + h$ from an approximation at time t. This allows the computation of approximate values $y_1, y_2, y_3, \ldots$ of the solution at a sequence of times $t_1, t_2, t_3, \ldots$, where $t_n = t_0 + n\,h$. The simplest such method is known as **Euler's Method**:

$$\text{Given } t_0 \text{ and } y_0, \text{ compute } \begin{cases} t_{n+1} = t_n + h \\ y_{n+1} = y_n + h\,f(t_n, y_n) \end{cases} \text{ for } n = 1, 2, 3, \ldots$$

The parameter h is called the *stepsize*. It is typically some small, positive number such as 0.05 or 0.01. Roughly speaking, the smaller the stepsize h is, the better the approximation will be. However, smaller values of h require more steps to reach any given value of t.

Euler's Method is simple to implement in *Mathematica*. First we define a function that computes each successive point (t, y) on the graph of the approximate solution from the previous one:

```
eulerStep[{t_, y_}] := {t, y} + h {1, f[t, y]}
```

Then, after defining $f(x)$ and h, we use `NestList` to generate the resulting list of ordered pairs (t_n, y_n) and use `ListPlot` to plot the resulting approximation.

■ Example 6.3.1

Plot an approximate solution of

$$\frac{dy}{dt} = \sin(t\,y), \quad y(0) = 2$$

on the interval $0 \le t \le 5$, *using Euler's method with stepsize* $h = 0.1$.

After defining

```
f[t_, y_] := Sin[t y]
h := 0.1
```

we use `NestList` as follows to compute a list of points on the approximate solution. Note that the second argument represents the starting point $(0, 2)$ and that fifty iterations are needed to reach $t = 5$.

```
points = NestList[eulerStep, {0, 2}, 50]

{{0, 2}, {0.1, 2}, {0.2, 2.01987}, {0.3, 2.05917},
 {0.4, 2.11709}, {0.5, 2.19201}, {0.6, 2.28095}, {0.7, 2.37891},
 {0.8, 2.47847}, {0.9, 2.5701}, {1., 2.64379}, {1.1, 2.69154},
 {1.2, 2.70953}, {1.3, 2.69857}, {1.4, 2.66273}, {1.5, 2.60741},
 {1.6, 2.53783}, {1.7, 2.45833}, {1.8, 2.37222}, {1.9, 2.28184},
 {2., 2.18886}, {2.1, 2.09441}, {2.2, 1.9993}, {2.3, 1.90419},
 {2.4, 1.80968}, {2.5, 1.71641}, {2.6, 1.62516}, {2.7, 1.53678},
 {2.8, 1.45222}, {2.9, 1.37238}, {3., 1.29803}, {3.1, 1.22968},
 {3.2, 1.16755}, {3.3, 1.11154}, {3.4, 1.06129}, {3.5, 1.01629},
 {3.6, 0.975929}, {3.7, 0.939604}, {3.8, 0.906733}, {3.9, 0.8768},
 {4., 0.849363}, {4.1, 0.824056}, {4.2, 0.800573}, {4.3, 0.778671},
 {4.4, 0.758149}, {4.5, 0.738844}, {4.6, 0.720626},
 {4.7, 0.703384}, {4.8, 0.687027}, {4.9, 0.671476}, {5., 0.656667}}
```

These points can now be plotted with `ListPlot`:

```
ListPlot[points]
```

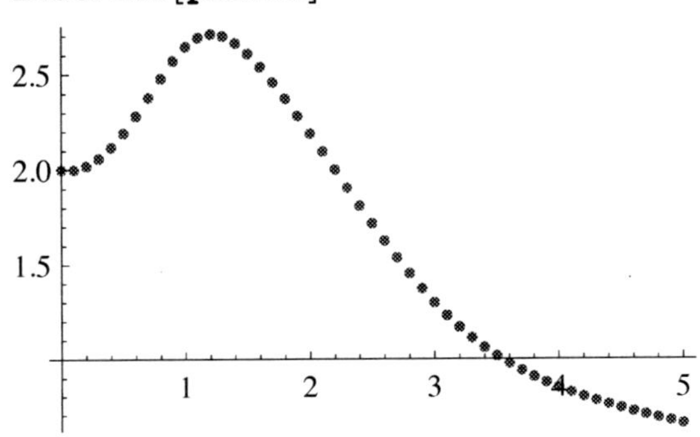

To join the points with line segments to form a nearly smooth curve, we can specify the option `PlotJoined→True`. (The option `PlotRange→{0,3}` causes the axes to intersect at the origin.)

```
ListPlot[points, PlotJoined → True, PlotRange → {0, 3}]
```

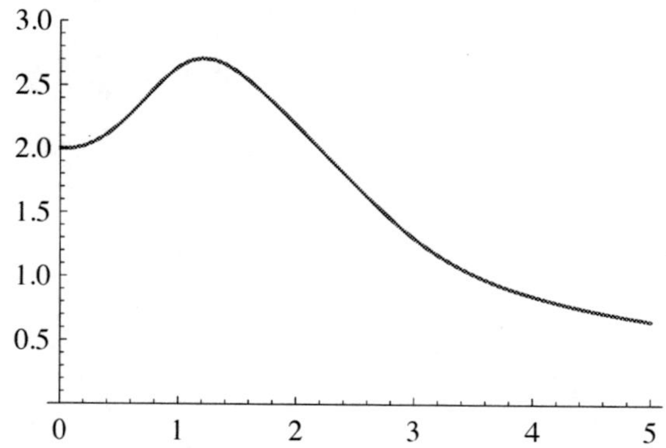

■ Example 6.3.2

Plot an approximate solution of

$$\frac{dy}{dt} = \cos y - \sin t, \quad y(0) = 1$$

on the interval $0 \le t \le 10$, *along with the direction field for the differential equation. Use Euler's method with stepsize* $h = 0.1$.

After defining

```
f[t_, y_] := Cos[y] - Sin[t]
h := 0.1
```

we use `NestList` as follows to compute a list of points on the approximate solution. Note that the second argument represents the starting point (0, 1) and that 100 iterations are needed to reach $t = 10$. (The semicolon at the end suppresses the lengthy output.)

```
points = NestList[eulerStep, {0, 1}, 100];
```

We now plot a graph through these points with `ListPlot`:

```
curve = ListPlot[points, PlotJoined → True,
  PlotStyle → Thickness[Medium], AspectRatio → Automatic]
```

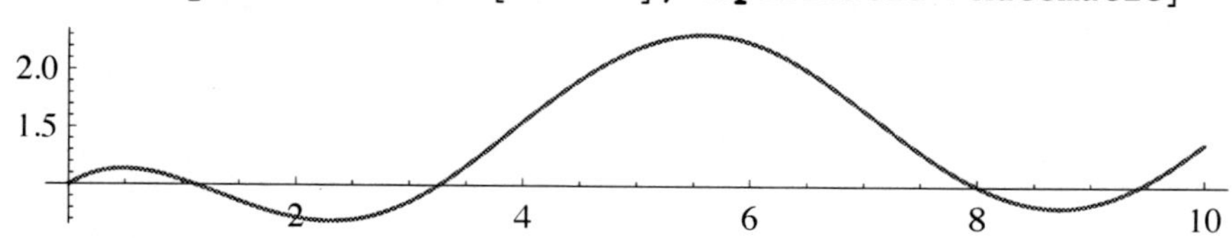

Next we'll use `plotDirectionField` from the previous section to plot the direction field.

```
field = plotDirectionField[f[t, y], {t, 0, 10}, {y, 0, 2.5}]
```

Finally now we'll superimpose the curve and the direction field.

```
Show[field, curve, Axes → True]
```

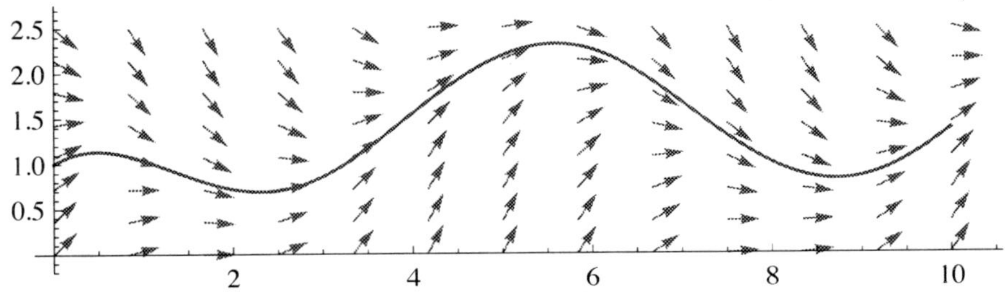

◆ Exercises

1. The exact solution of $\frac{dy}{dt} = t^2 - y$, $y(0) = 2$, is $y(t) = 2 - 2t + t^2$. Plot the exact solution along with Euler's Method approximations to the solution for $0 \le t \le 3$, using:

   a) 3 steps with stepsize $h = 1$
   b) 6 steps with stepsize $h = .5$
   c) 12 steps with stepsize $h = .25$
   d) 30 steps with stepsize $h = .1$
   e) 60 steps with stepsize $h = .05$

   In Exercises 2-5, use Euler's method with the suggested stepsize to plot an approximate solution on the indicated interval.

2. $\frac{dy}{dt} = \left(e^{2-y} - 1\right) y, \quad y(0) = .1\,; \quad h = 0.1, \quad 0 \le t \le 2$

3. $\frac{dy}{dt} = \sin\left(t^2 + y\right), \quad y(0) = 0\,; \quad h = 0.05, \quad 0 \le t \le 5$

4. $\frac{dy}{dt} = \sin t^2, \quad y(0) = 0\,; \quad h = 0.05, \quad 0 \le t \le 5$

5. $\frac{dy}{dt} = \frac{1}{2}\, y\,(3 - 2\sin(2\pi t) - y), \quad y(0) = 1\,; \quad h = 0.05, \quad 0 \le t \le 5$

6.4 Exact Solutions

There are several types of differential equations for which there are standard methods for finding exact solutions. In this section we will look at methods for solving *separable* equations and *first-order linear* equations.

■ Separable Equations

A first-order differential equation that can be written in the form

$$f(y)\,dy = g(t)\,dt$$

is said to be separable. With variables separated in this way, the differentials on each side of the equation may be antidifferentiated independently:

$$\int f(y)\,dy = \int g(t)\,dt$$

In some cases, the resulting equation can be solved explicitly for y in terms of t.

■ Example 6.4.1

Find the general solution of $\frac{dy}{dt} = y^2 \sin t$.

The separated form of this equation is $y^{-2}\,dy = \sin t\;dt$. Antidifferentiating each side of this separated equation produces an algebraic equation involving y, t, and an arbitrary constant c. Note that since *Mathematica* does not give an arbitrary constant with its antiderivatives, we must introduce c into the equation ourselves.

```
algEq = ∫y^-2 ⅆy == ∫Sin[t] ⅆt + c
```

```
  1
- — == c - Cos[t]
  y
```

Solving for y now gives us the general solution.

```
y /. First@Solve[algEq, y]
```

```
     1
———————————
-c + Cos[t]
```

■ Example 6.4.2

Find and graph the solution of the initial value problem

$$\frac{dy}{dt} = -t\sqrt{y}\,, \quad y(0) = 3.$$

The separated form of this equation is $y^{-1/2}\, dy = -t\, dt$. As in Example 6.4.1, we antidifferentiate each side of the separated equation, adding an arbitrary constant to one side.

```
algEq = ∫y^(-1/2) dy == ∫-t dt + c
```

$$2\sqrt{y} == c - \frac{t^2}{2}$$

At this point we substitute $t = 0$ and $y = 3$, solve for c, and then replace c in the equation with the resulting value.

```
algEq /. t → 0 /. y → 3
Solve[%, c] // Flatten
algEq = algEq /. %
```

$$2\sqrt{3} == c$$

$$\{c \to 2\sqrt{3}\}$$

$$2\sqrt{y} == 2\sqrt{3} - \frac{t^2}{2}$$

Now we solve for y and graph the solution.

```
y[t_] = y /. Solve[algEq, y][[1]]
```

$$\frac{1}{16}\left(48 - 8\sqrt{3}\, t^2 + t^4\right)$$

```
Plot[y[t], {t, 0, 4}]
```

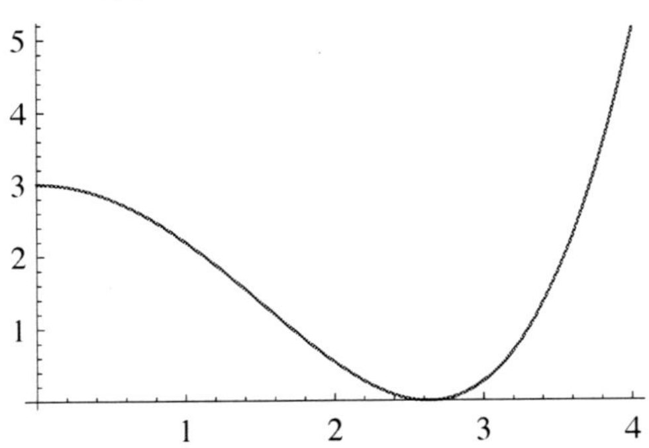


```
Clear[y]
```

■ Integration Between Limits

A very useful technique for solving initial value problems is definite—rather than indefinite— integration, i.e., *integration between limits*. Given a separable differential equation together with an initial value:

$$f(y)\, dy = g(t)\, dt, \quad y(t_0) = y_0,$$

we can find the solution by introducing dummy variables of integration and integ- rating the left side of the equation from y_0 to y and the right side from t_0 to t:

$$\int_{y_0}^{y} f(u)\, du = \int_{t_0}^{t} g(s)\, ds.$$

We will illustrate this technique with an example of the **logistic equation**:

$$\frac{dy}{dt} = k\left(1 - \frac{y}{M}\right) y.$$

■ Example 6.4.3

Find and plot the solution of $\frac{dy}{dt} = \frac{1}{10}\left(1 - \frac{y}{100}\right) y$, $y(0) = 5$.

One separated form of the equation is

$$\frac{100\, dy}{(100 - y)\, y} = \frac{dt}{10}.$$

Using the integration-between-limits technique described above, we obtain

```
algEq = Assuming[5 < y < 100 && t > 0, ∫_5^y 100/((100 - u) u) du == ∫_0^t 1/10 ds]
```

$$\text{Log}\left[-\frac{19\,y}{-100 + y}\right] == \frac{t}{10}$$

Now we solve for y and plot the graph:

```
y[t_] = y /. First[Solve[algEq, y]]
```

$$\frac{100\, e^{t/10}}{19 + e^{t/10}}$$

```
Plot[y[t], {t, -20, 100}]
```

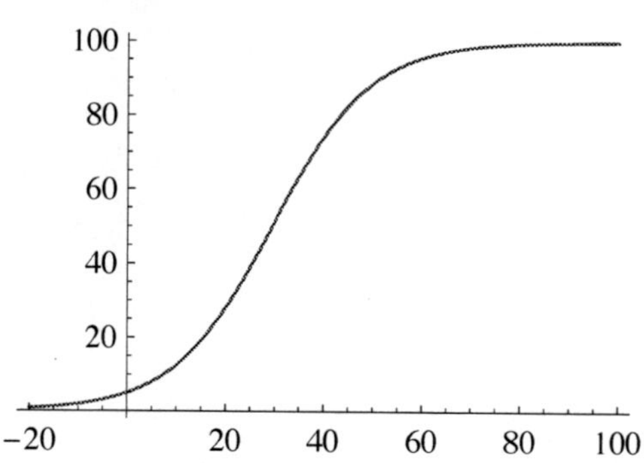


```
Clear[y]
```

Linear First-order Equations

Recall that linear first-order equations are of the form

$$\frac{dy}{dt} + p(t)\, y = q(t).$$

When $p(t)$ and $q(t)$ are constant functions, such an equation is separable.

Example 6.4.4

Find and plot the solution of

$$\frac{dy}{dt} + 5\, y = 10, \quad y(0) = 5.$$

To put the equation in separated form, we first rewrite it as $\frac{dy}{dt} = 10 - 5\, y$. Then we see that one separated form of the equation is

$$\frac{dy}{2-y} = 5\, dt.$$

Using the integration-between-limits technique described above, we obtain

```
algEq = Assuming[2 < y < 5 && t > 0, ∫_5^y 1 / (2 - u) du == ∫_0^t 5 ds]
```

$$\text{Log}\left[\frac{3}{-2+y}\right] == 5\, t$$

```
y[t_] = y /. Solve[algEq, y][[1]]
```

$$e^{-5t}\left(3 + 2\, e^{5t}\right)$$

```
Plot[y[t], {t, 0, 1}, PlotRange → All]
```

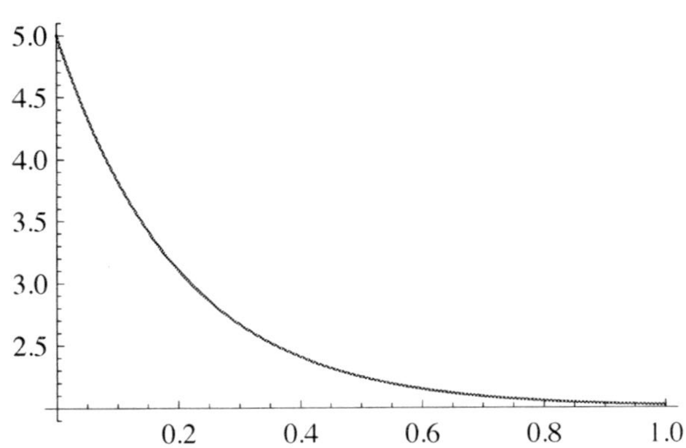


```
Clear[y]
```

When a first-order linear equation is not separable, the key to its solution is to multiply through by an appropriate **integrating factor**. The purpose of the integrating factor is to make the left side of the equation recognizable as the derivative of a product. The integrating factor that serves this purpose for

$$\frac{dy}{dt} + p(t)\,y = q(t)$$

is

$$e^{\int p},$$

where the notation $\int p$ means any antiderivative of $p(t)$. Multiplying each side of the equation by $e^{\int p}$ gives

$$e^{\int p}\frac{dy}{dt} + p(t)\,e^{\int p}\,y = e^{\int p}\,q(t),$$

which, because of the product rule, can be rewritten as

$$\frac{d}{dt}\left(e^{\int p}\,y\right) = e^{\int p}\,q(t),$$

With this done, we can now find y by simply antidifferentiating each side of the equation and then multiplying through by $e^{-\int p}$.

▪ Example 6.4.5

Find the general solution of $\frac{dy}{dt} + y = \sin t$.

Here we have $p(t) = 1$ and $q(t) = \sin t$. The integrating factor is

$$e^{\int 1} = e^{t}.$$

Multiplying through the equation by e^t and recognizing the left side as the derivative of a product gives us

$$\frac{d}{dt}\left(e^{t}\,y\right) = e^{t}\sin t,$$

which implies that

```
ⅇ^t y == ∫ ⅇ^t Sin[t] ⅆt + c
```

```
ⅇ^t y == c + 1/2 ⅇ^t (-Cos[t] + Sin[t])
```

Now multiplying through by e^{-t} gives the general solution:

```
Thread[ⅇ^-t %, Equal] // Simplify
```

```
2 y + Cos[t] == 2 c ⅇ^-t + Sin[t]
```

An integration-between-limits technique can be used to solve initial value problems that involve first-order linear equations. Suppose we have an initial value problem

$$\frac{dy}{dt} + p(t)\,y = q(t), \quad y(t_0) = y_0.$$

The integrating factor technique described above gives

$$\frac{d}{dt}\left(e^{\int p}\,y\right) = e^{\int p}\,q(t)\,.$$

To simplify notation, let $\phi(t) = e^{\int p}$. Then after multiplying through by dt, we have

$$d(\phi(t)\,y) = \phi(t)\,q(t)\,dt\,.$$

Now we replace t with a different dummy variable and then integrate each side of this equation from t_0 to t to produce

$$\phi(t)\,y(t) - \phi(t_0)\,y_0 = \int_{t_0}^{t} \phi(s)\,q(s)\,ds.$$

Thus we arrive at the solution

$$y(t) = \phi(t)^{-1}\left(\phi(t_0)\,y_0 + \int_{t_0}^{t} \phi(s)\,q(s)\,ds\right).$$

■ Example 6.4.6

Find, for $t \le 5$, the solution of

$$\frac{dy}{dt} + \frac{3\,y}{5-t} = t, \quad y(0) = y_0.$$

Plot the graph of the solution for $0 \le t < 5$ with $y_0 = 10$.

First we compute the integrating factor $\phi(t)$.

```
ϕ[t_] = e^∫3/(5-t) dt
```

Now we'll simply enter the formula derived above.

```
y[t_] = ϕ[t]^-1 (ϕ[0] y0 + ∫_0^t ϕ[s] s ds) // Simplify
```

The solution with $y_0 = 10$ is now obtained and plotted as follows.

```
y[t_] = y[t] /. y0 → 10

Plot[y[t], {t, 0, 5}]; Clear[y, ϕ];
```

■ Example 6.4.7

The velocity $v(t)$ of a free-falling object, under the influence of constant gravitational force *and air resistance*, can be modeled by

$$\frac{dv}{dt} + \frac{k}{m} v = -g, \quad v(0) = v_0.$$

We assume that k, m, and g are constants. *Find the solution in terms of k, m, g, and v_0. Then plot the solution for $k = m = 1$, $g = 32$, and $v_0 = 0$.*

The integrating factor here is

```
ϕ[t_] = e^∫k/m dt
```

$$e^{\frac{k\,t}{m}}$$

So the solution is

```
v[t_] = ϕ[t]^-1 (ϕ[0] v0 + ∫_0^t ϕ[s] (-g) ds) // Simplify
```

$$e^{-\frac{k\,t}{m}}\left(-\frac{\left(-1+e^{\frac{k\,t}{m}}\right) g\, m}{k} + v0\right)$$

Now we substitute in the given parameters and plot the result:

```
v[t_] = v[t] /. {k → 1, m → 1, g → 32, v0 → 0}
```

$$-32\, e^{-t}\left(-1 + e^{t}\right)$$

```
Plot[v[t], {t, 0, 5}]
```

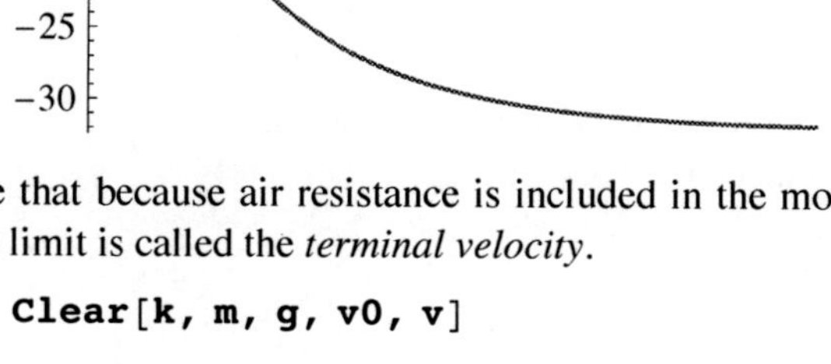

Note that because air resistance is included in the model, the velocity has a limit as $t \to \infty$. This limit is called the *terminal velocity*.

```
Clear[k, m, g, v0, v]
```

◆ Exercises

In Exercises 1-6, find the general solution of the differential equation.

1. $\dfrac{dy}{dt} = -\sqrt{y}$
2. $\dfrac{dy}{dt} + y\cos t = \cos t$
3. $\dfrac{dy}{dt} + \dfrac{\cos t}{2 + \sin t}\, y = \sin t$
4. $\dfrac{dy}{dt} = y\,(\ln y)^2$
5. $\dfrac{dy}{dt} - y = \sin t$
6. $\dfrac{dy}{dt} = (1 + y)^2\left(1 - t^2\right)$

7.-12. For each of the equations in exercises 1-6, use the integration between limits technique (as in Example 6.4.3 or Examples 6.4.6-7 as appropriate) to find the solution satisfying the initial condition $y(0) = y_0$. For the equation in Exercise 4, assume that $y_0 > 0$.

13. Use the velocity function $v(t)$ derived in Example 6.4.7 in terms of k, m, g, and v_0 to compute the height function

$$y(t) = y_0 + \int_0^t v(s)\, ds.$$

Then take $k = m = 1$, $g = 32$, $v_0 = 0$, and $y_0 = 200$, and plot $v(t)$ and $y(t)$ for $0 \le t \le t^*$ where $y(t^*) = 0$.

6.5 DSolve and NDSolve

Mathematica has built-in functions for solving differential equations. `DSolve` finds exact solutions and `NDSolve` computes numerical approximations.

■ Using DSolve

■ Example 6.5.1

Use `DSolve` *to find the general solution of* $\dfrac{dy}{dt} + y = \cos t$.

This illustrates the basic usage of `DSolve`:

```
Clear[y]

DSolve[y'[t] + y[t] == Cos[t], y[t], t]

{{y[t] → e^-t C[1] + 1/2 (Cos[t] + Sin[t])}}
```

Notice that, just as with `Solve` and other *Mathematica* functions we have seen, the result is given as a list of one or more replacement rules. Also note that the arbitrary constant is represented by `C[1]`.

■ Example 6.5.2

Use `DSolve` *to find the general solution of the* second-order *equation*

$$\frac{d^2y}{dt^2} + 2\frac{dy}{dt} + 2y = t.$$

This is the result:

```
DSolve[y''[t] + 2 y'[t] + 2 y[t] == t, y[t], t]
```

$$\left\{\left\{y[t] \to \frac{1}{2}(-1+t) + e^{-t}\,C[2]\,\text{Cos}[t] + e^{-t}\,C[1]\,\text{Sin}[t]\right\}\right\}$$

Notice that there are two arbitrary constants, `C[1]` and `C[2]`.

■ Example 6.5.3

Use `DSolve` *to find the solution of the initial value problem*

$$\frac{dy}{dt} + y = e^{-2t}, \quad y(0) = y_0.$$

To solve an initial value problem, we simply include the initial condition as a second equation in a list:

```
y[t] /. First@DSolve[{y'[t] + y[t] == e^-2 t, y[0] == y0}, y[t], t]
```

$$e^{-2t}\left(-1 + e^{t} + e^{t}\,y0\right)$$

■ Example 6.5.4

Use `DSolve` *to find the solution of*

$$\frac{dy}{dt} + t\,y = y^2, \quad y(0) = y_0.$$

This time `DSolve` returns the solution in terms of the special function `Erf`.

```
y[t] /. First@DSolve[{y'[t] + t y[t] == y[t]^2, y[0] == y0}, y[t], t]
```

$$-\frac{2\,e^{-\frac{t^2}{2}}\,y0}{-2 + \sqrt{2\pi}\,y0\,\text{Erf}\left[\frac{t}{\sqrt{2}}\right]}$$

■ Example 6.5.5

Use `DSolve` *to find the solution of*

$$\frac{dy}{dt} = t^2 - y^3, \quad y(0) = 1.$$

This time `DSolve` fails to find the solution.

```
DSolve[{y'[t] == t^2 - y[t]^3, y[0] == 1}, y[t], t]
```

```
DSolve[{y'[t] == t^2 - y[t]^3, y[0] == 1}, y[t], t]
```

Using NDSolve

Example 6.5.6

Use `NDSolve` *to find and plot an approximate solution of*

$$\frac{dy}{dt} = t^2 - y^3, \quad y(0) = 2$$

on the interval $0 \le t \le 5$.

`NDSolve` resurns its result in the form of an `InterpolatingFunction`:

```
numSol = NDSolve[{y'[t] == t^2 - y[t]^3, y[0] == 2}, y, {t, 0, 5}]
{{y → InterpolatingFunction[{{0., 5.}}, <>]}}
```

This defines a function that evaluates the `InterpolatingFunction`:

```
yApprox[t_] = y[t] /. First[numSol]
InterpolatingFunction[{{0., 5.}}, <>][t]
```

Now it is easy to plot the result.

```
Plot[yApprox[t], {t, 0, 5}]
```

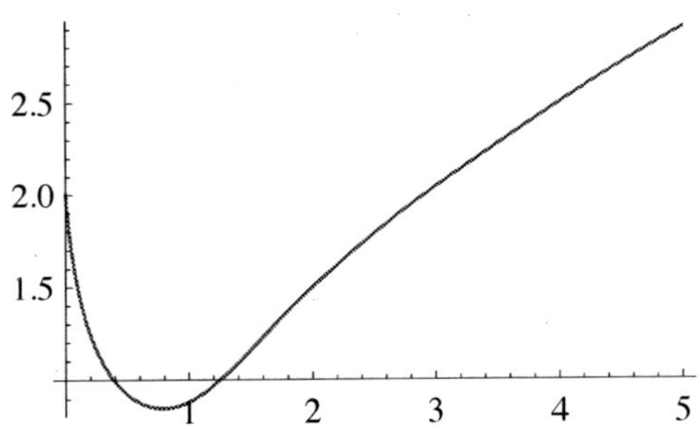

Example 6.5.7

Use `NDSolve` *to find and plot an approximate solution of*

$$\frac{dy}{dt} = 3\cos(t\,y), \quad y(0) = y_0$$

on the interval $0 \le t \le 5$ *for each of the initial values* $y_0 = 0, 1, 2, \ldots, 5$.

Let's start by using `Table` to create a list containing the output of `NDSolve` for each initial value:

```
numSolns = Table[NDSolve[
   {y'[t] == 3 Cos[t y[t]], y[0] == y0}, y, {t, 0, 5}], {y0, 0, 5, 1}]
{{{y → InterpolatingFunction[{{0., 5.}}, <>]}},
 {{y → InterpolatingFunction[{{0., 5.}}, <>]}},
 {{y → InterpolatingFunction[{{0., 5.}}, <>]}},
 {{y → InterpolatingFunction[{{0., 5.}}, <>]}},
 {{y → InterpolatingFunction[{{0., 5.}}, <>]}},
 {{y → InterpolatingFunction[{{0., 5.}}, <>]}}}
```

Now we'll convert that into a list of functions:

```
yApprox[t_] = y[t] /. numSolns // Flatten
```

```
{InterpolatingFunction[{{0., 5.}}, <>][t],
 InterpolatingFunction[{{0., 5.}}, <>][t],
 InterpolatingFunction[{{0., 5.}}, <>][t],
 InterpolatingFunction[{{0., 5.}}, <>][t],
 InterpolatingFunction[{{0., 5.}}, <>][t],
 InterpolatingFunction[{{0., 5.}}, <>][t]}
```

Now we plot the result (with the help of `Evaluate`):

```
Plot[Evaluate[yApprox[t]], {t, 0, 5}]
```

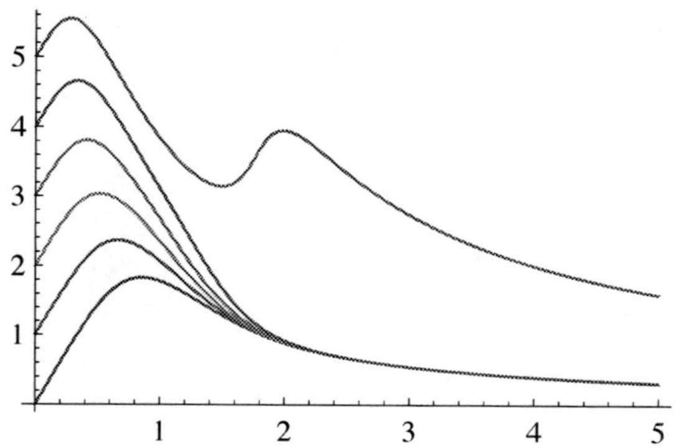

■ Exercises

1.-12. Rework Exercises 1-12 in Section 6.4 using `DSolve`.

In Exercises 13-16, use `NDSolve` to find and plot approximate solutions on the indicated interval for each given initial value. Also `Show` the plot together with the direction field for the differential equation. (See Section 6.2.)

13. $\frac{dy}{dt} = \left(e^{2-y} - 1\right) y, \quad y(0) = .5, 1, 1.5, \ldots, 4; \quad 0 \le t \le 3$

14. $\frac{dy}{dt} = \sin\left(t^2 + y^2\right), \quad y(0) = -1, -.75, -.5, \ldots, 1; \quad 0 \le t \le 7$

15. $\frac{dy}{dt} = \sin t^2, \quad y(0) = 0, .5, 1, 1.5, 2; \quad 0 \le t \le 5$

16. $\frac{dy}{dt} = \frac{1}{2} y (3 - 2\sin(2\pi t - y)), \quad y(0) = 1, 2, 3, 4, 5; \quad 0 \le t \le 4$

6.6 Systems of Differential Equations

We consider here initial value problems involving pairs of coupled, *autonomous*, first-order differential equations. These are of the form

$$\frac{dx}{dt} = f(x, y), \quad x(0) = x_0$$
$$\frac{dy}{dt} = g(x, y), \quad y(0) = y_0$$

where f and g are given continuous functions of two variables. (The differential equations are called autonomous because f and g depend only upon x and y—not upon t.) Think of the solution of such a problem as a pair of parametric equations

$$x = x(t), \;\; y = y(t),$$

whose graph is (naturally) a parametric curve.

■ Example 6.6.1

Show that $x = 2 \sin t + \cos t$ *and* $y = \sin t$ *satisfy the system*

$$x' = 2x - 5y, \quad x(0) = 1$$
$$y' = x - 2y, \quad y(0) = 0$$

and plot the corresponding parametric curve.

Let's first enter the functions.

```
x[t_] := 2 Sin[t] + Cos[t];  y[t_] := Sin[t]
```

This shows that the x-equation and the initial value for x are satisfied:

```
x'[t] - 2 x[t] + 5 y[t] // Simplify
x[0]
0
1
```

This shows that the y-equation and the initial value for y are satisfied:

```
y'[t] - x[t] + 2 y[t] // Simplify
y[0]
0
0
```

Now let's plot the curve.

```
ellipse = ParametricPlot[{x[t], y[t]}, {t, 0, 2 π}]
```

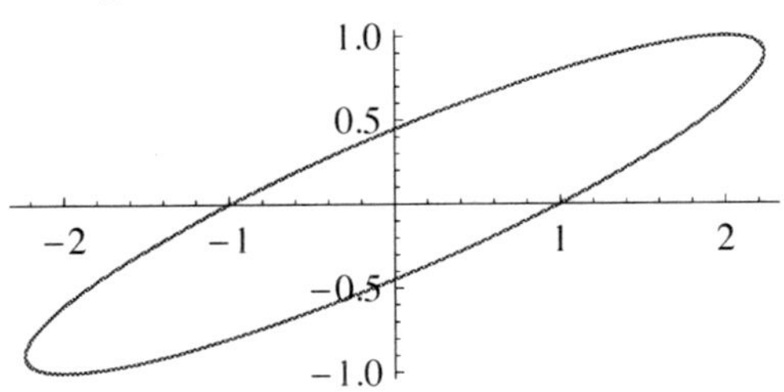

■ Direction Fields

The `VectorFieldPlot` function that we used in Section 6.2 can be used to plot the direction field for a pair of differential equations. For the system in Example 6.6.1 above, we can plot the direction field as follows. Note that the first argument of `VectorFieldPlot` is the list $\{f(x, y), g(x, y)\}$.

```
Needs["VectorFieldPlots`"]
```

```
field = VectorFieldPlot[{2 x - 5 y, x - 2 y},
  {x, -2.5, 2.5}, {y, -1.5, 1.5}, PlotPoints → 11,
  ScaleFunction → (1 &), Axes → True, ColorFunction → (Blue &)]
```

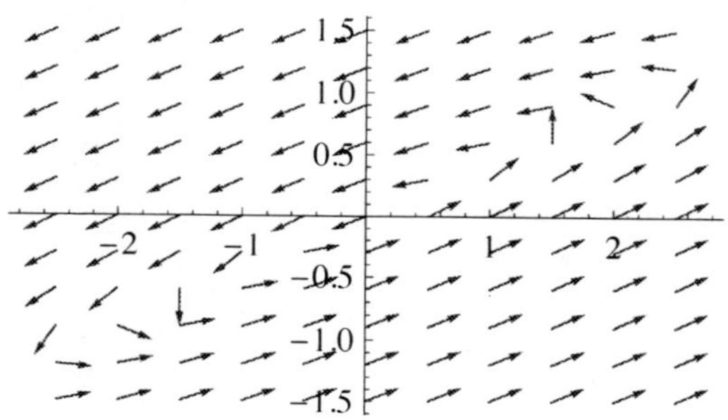

▪ Example 6.6.2

Plot the direction field for the system

$$\frac{dx}{dt} = -x - y, \quad \frac{dy}{dt} = 2x - y.$$

Infer from the direction field the behavior of solutions of the system.

We plot the direction field as follows.

```
field = VectorFieldPlot[{-x - y, 2 x - y},
  {x, -3, 3}, {y, -2, 2}, PlotPoints → {13, 7},
  ScaleFunction → (1 &), ColorFunction → (Red &)]
```

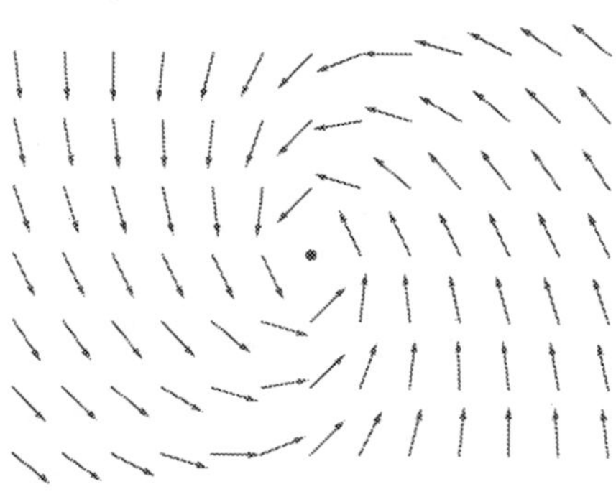

It is apparent from the direction field that graphs of solutions curves of the system spiral in toward the origin as $t \to \infty$. This implies that $x(t)$ and $y(t)$ have an oscillatory nature with amplitude that decays to zero as $t \to \infty$.

■ Example 6.6.3

Plot the direction field for the system

$$\frac{dx}{dt} = x + y, \quad \frac{dy}{dt} = x - y.$$

Infer from the direction field the behavior of solutions of the system.

The direction field is plotted as follows.

```
field = VectorFieldPlot[
  {x + y, x - y}, {x, -3, 3}, {y, -2, 2}, PlotPoints → {13, 7},
  ScaleFunction → (1 &), ColorFunction → (Black &)]
```

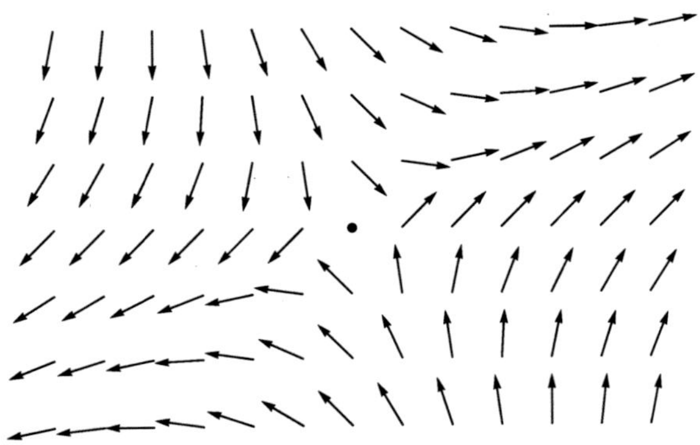

It is apparent from the direction field that graphs of solutions of the system have a *hyperbolic* character—approaching the origin but eventually veering away from it. In particular, it *appears* that for any given initial values, $x(t)$ and $y(t)$ approach either $\pm\infty$ as $t \to \infty$. (See Exercise 3.)

■ Euler's Method

The solution of a system

$$\frac{dx}{dt} = f(x, y), \quad x(0) = x_0$$

$$\frac{dy}{dt} = g(x, y), \quad y(0) = y_0$$

can be approximated by Euler's Method, which in this case takes the form:

$$\text{Given } t_0 \text{ and } y_0, \text{ compute } \begin{cases} x_{n+1} &= x_n + h\, f(x_n, y_n) \\ y_{n+1} &= y_n + h\, g(x_n, y_n) \end{cases} \quad \text{for } n = 1, 2, 3, \ldots$$

This is nearly as simple to implement in *Mathematica* as was the single equation case. The following function computes one step of the method.

```
eulerStep[{x_, y_}] := {x, y} + h {f[x, y], g[x, y]}
```

■ Example 6.6.4

Use Euler's Method with $h = 0.05$ to compute an approximate solution of the system

$$\frac{dx}{dt} = -x - y, \quad x(0) = 3$$

$$\frac{dy}{dt} = 2x - y, \quad y(0) = -2$$

for $0 \le t \le 5$. Plot the result with `ListPlot` *and then show the result on top of the direction field plotted in Example 6.6.2.*

We begin by entering definitions for the functions f and g.

```
f[x_, y_] := -x - y; g[x_, y_] := 2 x - y
```

Since we are using $h = .05$, we need to compute 100 steps to reach $t = 5$. Using `NestList`, with starting point $(3, -2)$, 100 points are computed by entering

```
h := .05; points = NestList[eulerStep, {3, -2}, 100];
```

Note that the semicolon at the end suppresses what would be a very large output. Having computed these points, we plot an approximate solution curve with `ListPlot`.

```
curve = ListPlot[points, PlotJoined → True]
```

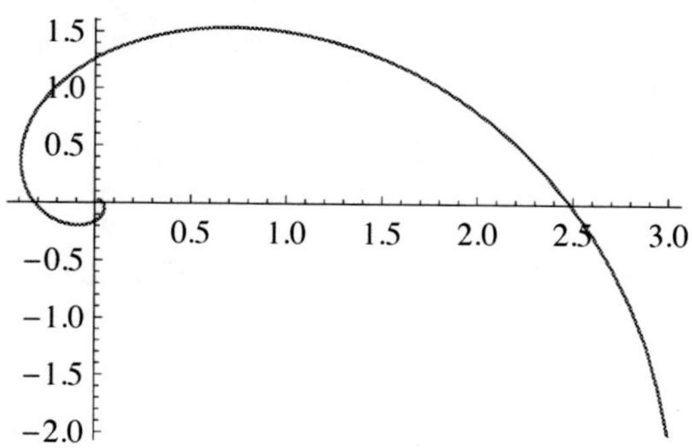

Finally, using the direction field from Example 6.6.2, we have the following plot.

```
Show[curve, field, PlotRange → All]
```

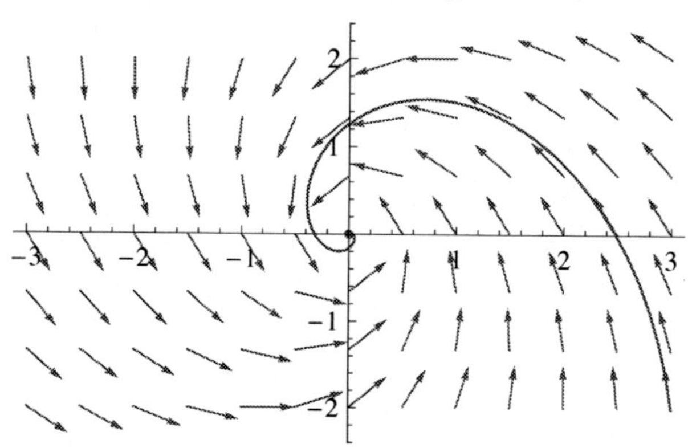

■ NDSolve

As you might imagine, finding exact solutions of coupled pairs of differential equations is a complicated task at best. For autonomous *linear* systems, i.e., systems in which the functions f and g are linear, it is always possible to derive fairly simple exact solutions in terms of exponential and trigonometric functions. For nonlinear systems, finding exact solutions is rarely possible.

Euler's Method is a useful and simple method for computing approximate solutions; yet unless we use a *very* small stepsize, it lacks the accuracy needed to produce approximate solutions that reflect the true behavior of many interesting systems, particularly nonlinear ones. There are many methods that are far more accurate than Euler's Method and can be made *adaptive* in the sense that the stepsize is automatically adjusted to maintain and guarantee a certain degree of accuracy. *Mathematica*'s `NDSolve` uses such advanced methods to compute highly accurate approximate solutions.

▪ Example 6.6.5

Using Euler's Method with $h = 0.01$, *compute and plot, for* $0 \le t \le 7$, *an approximate solution of the system*

$$x' = -y, \quad x(0) = 1$$

$$y' = x, \quad y(0) = 0$$

(of which $x = \cos t$, $y = \sin t$ *is the exact solution). Then, for comparison, compute an approximate solution using* `NDSolve` *and plot the result.*

We'll use the `eulerStep` function we've already defined. With

```
f[x_, y_] := -y; g[x_, y_] := x
```

and $h = .01$, we will compute 700 steps to reach $t = 7$. (Note the use of the semicolon to suppress the output of `NestList`.)

```
h := .01
points = NestList[eulerStep, {1, 0}, 700];
```

Now we plot the result.

```
ListPlot[points, PlotJoined → True, AspectRatio → Automatic]
```

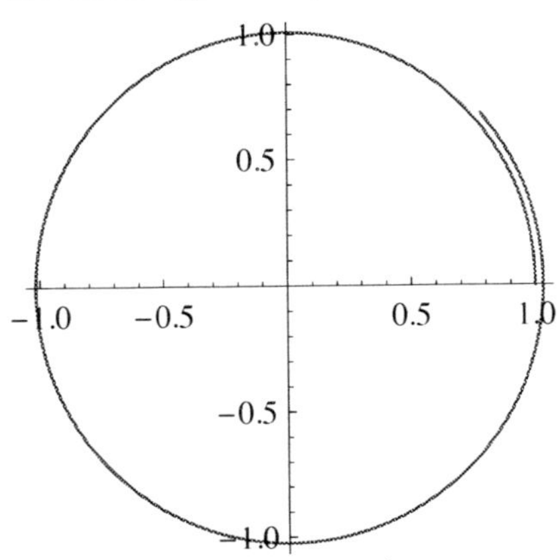

Notice that even though the stepsize was rather small, our approximation had accumulated a significant amount of error by the time the exact solution would have returned to the starting point (1, 0). In fact, the error at $t = 2\pi$ is given roughly by

```
{1, 0} - points[[628]]
```

```
{-0.0317524, 0.0138204}
```

Now let's use `NDSolve`. This computes the approximate solution:

```
Clear[x, y]
```

```
solns = {x[t], y[t]} /. First@
NDSolve[{x'[t] == -y[t], y'[t] == x[t], x[0] == 1, y[0] == 0},
    {x[t], y[t]}, {t, 0, 7}]
```

```
{InterpolatingFunction[{{0., 7.}}, <>][t],
 InterpolatingFunction[{{0., 7.}}, <>][t]}
```

We use `ParametricPlot` to plot the result.

```
ParametricPlot[Evaluate[solns],
 {t, 0, 7}, AspectRatio → Automatic]
```

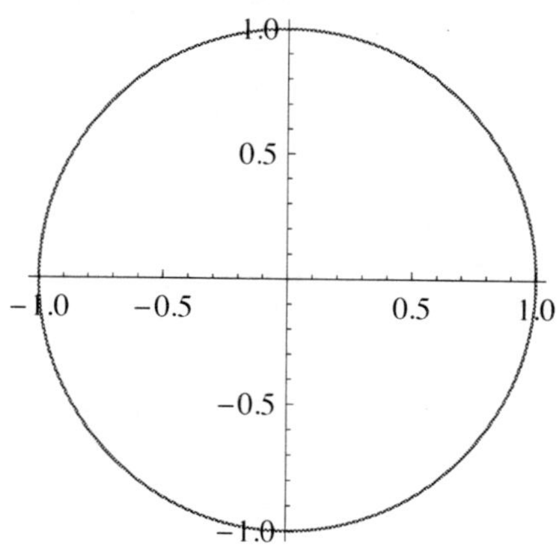

Notice that the result is more-or-less indistinguishable from the unit circle. In fact, the error at $t = 2\pi$ is given by

```
{1, 0} - solns /. t → 2 π
```

$\{7.81386 \times 10^{-7},\ 1.0643 \times 10^{-7}\}$

■ Example 6.6.6

Plot an approximate solution of the predator-prey *system*

$$x' = x\left(1 - \frac{y}{9}\right), \qquad x(0) = 4$$

$$y' = -y\left(1 - \frac{x}{12}\right), \quad y(0) = 4$$

for $0 \le t \le 10$, *together with the direction field.*

Let's start by defining functions f and g.

```
f[x_, y_] := x (1 - y/9); g[x_, y_] := -y (1 - x/12)
```

Then we'll use `NDSolve` to compute an approximate solution.

```
solns = {x[t], y[t]} /. First@
NDSolve[{x'[t] == f[x[t], y[t]], y'[t] == g[x[t], y[t]],
     x[0] == 4, y[0] == 4}, {x[t], y[t]}, {t, 0, 10}]

{InterpolatingFunction[{{0., 10.}}, <>][t],
 InterpolatingFunction[{{0., 10.}}, <>][t]}
```

This plots the approximate solution curve along with the direction field:

```
Show[ParametricPlot[Evaluate[solns], {t, 0, 10}],
 VectorFieldPlot[{f[x, y], g[x, y]}, {x, 0, 35}, {y, 0, 25},
  ScaleFunction → (1 &), ColorFunction → (Hue[0] &)], Axes → True]
```

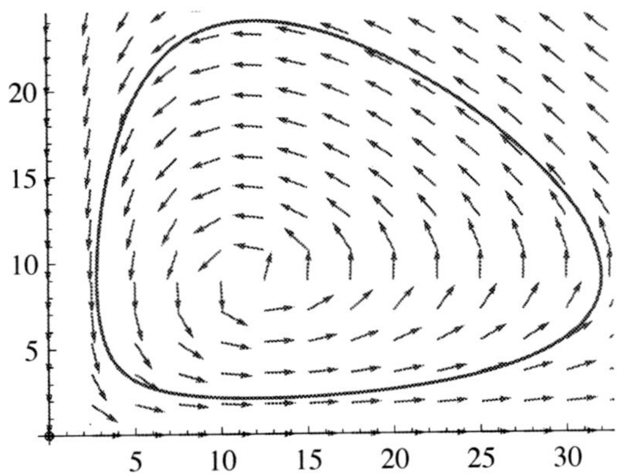

In this example $x(t)$ represents the prey population and $y(t)$ represents the predator population (each perhaps in units of hundreds or thousands).

◆ Exercises

1. Verify that $x = \sin 2t + \cos 2t$ and $y = \sin 2t - 3\cos 2t$ satisfy

$$x' = x + 5y, \quad y' = -x - y.$$

Plot this solution for $0 \le t \le \pi$, along with the direction field for the system.

2. Verify that $x = -e^t \sin 2t$ and $y = e^t \cos 2t$ satisfy

$$x' = x - 2y, \quad y' = 2x + y.$$

Plot this solution for $0 \le t \le \pi$, along with the direction field for the system.

3. Show that the system in Example 6.6.3,

$$x' = x + y, \quad y' = x - y,$$

has solutions satisfying $y(t) = m\,x(t)$ and that those solutions where $m < 0$ approach $(0, 0)$ as $t \to \infty$. (Hint: If $y(t) = m\,x(t)$, then $\dfrac{dy/dt}{dx/dt} = \dfrac{x - m\,x}{x + m\,x} = \dfrac{dy}{dx} = m$.)

4. Use Euler's Method with stepsize $h = .05$ to plot an approximate solution of the system

$$x' = y, \qquad x(0) = 0$$

$$y' = -x - y^3, \quad y(0) = 2$$

for $0 \leq t \leq 10$, along with the direction field for the system with $-2 \leq x \leq 2$ and $-2 \leq y \leq 2$. Compute another approximate solution using **`NDSolve`**. Plot it in color and then show both the Euler's Method curve and the **`NDSolve`** curve in the same plot (without the direction field).

5. The solution of the system in Example 6.6.6 is *periodic*, since its graph forms a "closed" curve. The period of the solution is the time it takes for the solution to return to its initial point. Determine the period of the solution, accurate to the nearest tenth.

6. For the solution of the system in Example 6.6.6, create a plot that shows the graphs of both $x(t)$ and $y(t)$ versus t. What do you observe? Can the same thing be seen in the parametric curve?

7. An example of one variation on the kind of predator-prey model in Example 6.6.6 is

$$\frac{dx}{dt} = x\left(1 - \frac{y}{9} - \frac{x}{25}\right), \quad x(0) = 20$$

$$\frac{dy}{dt} = -y\left(1 - \frac{x}{12}\right), \quad y(0) = 20.$$

Plot an approximate solution for $0 \leq t \leq 20$ along with the direction field for $0 \leq x \leq 25$ and $0 \leq y \leq 20$. How does the behavior of solutions of this system differ from that of the solutions of the system in Example 6.6.6?

7 Parametric and Polar Curves

In this chapter we will explore *Mathematica*'s ability to help us study parametric and polar curves.

7.1 Parametric Curves

A parametric curve in the plane is described by a pair of functions

$$x = f(t), \quad y = g(t)$$

where t varies over some interval. *Mathematica*'s `ParametricPlot` plots such curves.

■ Example 7.1.1: The Cycloid

A **cycloid** is a curve traced out by a fixed point on a circle as it rolls along a straight line. A circle with radius 1 traces out a cycloid parametrized by

$$x = t - \sin t, \quad y = 1 - \cos t .$$

```
cycloid = ParametricPlot[{t - Sin[t], 1 - Cos[t]}, {t, 0, 4 π}]
```

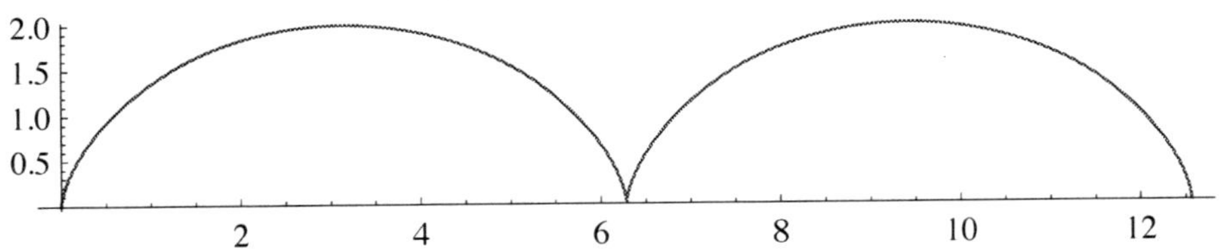

□ *Animating the Cycloid*

The following program creates an animation that shows two arches of a cycloid being drawn by a rolling wheel.

```
cycloid[] := Animate[Show[
   ParametricPlot[{t - Sin[t], 1 - Cos[t]}, {t, -3, k},
    PlotRange → {{-1, 13.5}, {0, 2.1}}, PlotStyle → Thick],
   Graphics[ {
     Table[Line[{{k, 1}, {k + Cos[θ - k], 1 + Sin[θ - k]}}],
      {θ, -3 π / 8, 11 π / 8, π / 8}], Thickness[Medium],
     Circle[{k, 1}, 1], PointSize[Large], Red,
     Line[{{k, 1}, {k + Cos[π / 2 + k], 1 - Sin[π / 2 + k]}}],
     Point[{k - Sin[k], 1 - Cos[k]}]} ], Ticks → None],
  {k, -2, 4 π + 2}]
```

```
cycloid[]
```

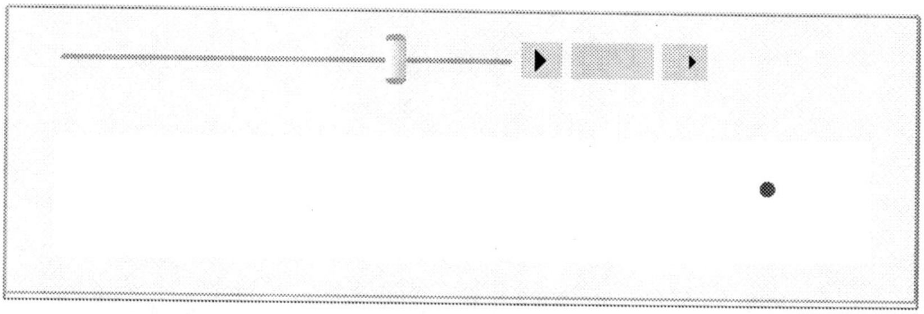

■ Example 7.1.2: Lissajous Figures

Lissajous figures are parametric curves of the form

$$x = a\cos t, \quad y = b\sin n\,t,$$

where $n > 1$. With $a = b = 1$ the Lissajous figures corresponding to $n = 2, 3, 4$ are plotted in a `GraphicsRow` as follows.

```
GraphicsRow[
 Table[ParametricPlot[{Cos[t], Sin[n t]},
   {t, 0, 2 π}, Ticks → {{-1, 1}, {-1, 1}}], {n, 2, 4}]]
```

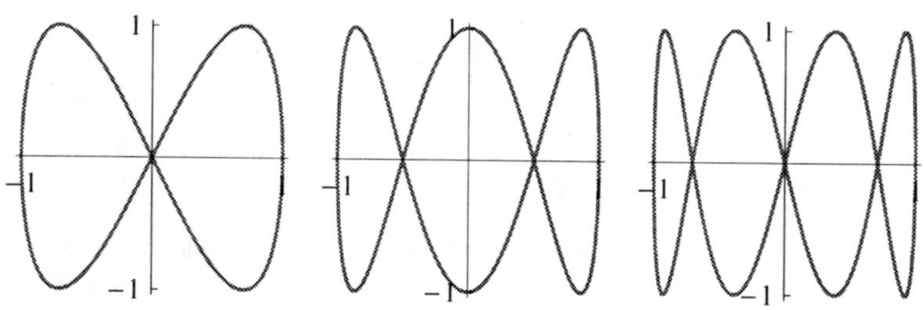

■ Example 7.1.3: Bézier Curves

Bézier curves are of central importance in the fields of modern typography and computer-aided design. A simple Bézier curve is determined by four "control points," which can be manipulated to predictably change the shape of the curve. The parametric equations of a simple Bézier curve with control points $(x_0, y_0), \ldots, (x_3, y_3)$ are the following cubic polynomials, where $0 \le t \le 1$.

$$x = x_0(1-t)^3 + 3\,x_1\,t\,(1-t)^2 + 3\,x_2\,t^2(1-t) + x_3\,t^3$$

$$y = y_0(1-t)^3 + 3\,y_1\,t\,(1-t)^2 + 3\,y_2\,t^2(1-t) + y_3\,t^3$$

Let's illustrate this with control points as follows:

```
controlPts := {{1, 0}, {4, 2}, {0, 2}, {3, 0}};
dots = Graphics[
   {PointSize[Medium], Darker[Red], Point /@ controlPts}];
Show[dots, Frame → True]
```

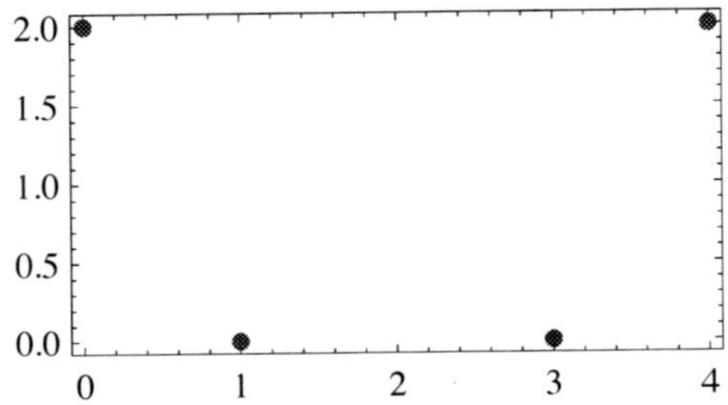

The Bézier curve associated with these points has the following parametrization.

```
b[t_] = {(1 - t)^3, 3 (1 - t)^2 t, 3 (1 - t) t^2, t^3}.controlPts
```

$\{(1-t)^3 + 12\,(1-t)^2\,t + 3\,t^3,\ 6\,(1-t)^2\,t + 6\,(1-t)\,t^2\}$

The resulting curve, together with the control points, is plotted as follows.

```
bezCurve = ParametricPlot[Evaluate[b[t]], {t, 0, 1}];
Show[dots, bezCurve, Frame → True]
```

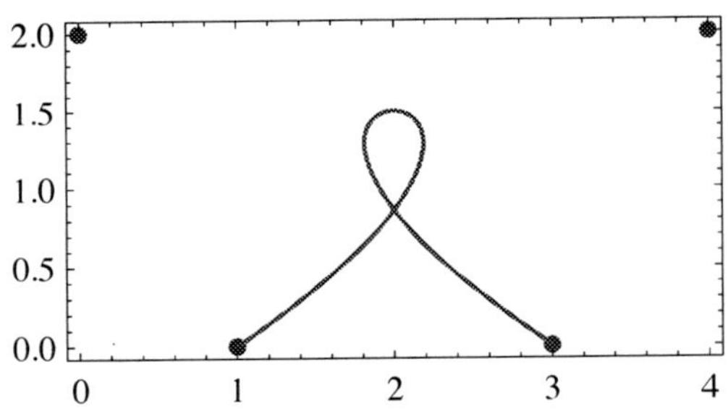

Notice that the curve starts at the first control point and ends at the fourth, a fact that is also seen by substituting 0 and 1 into the parametric equations of the curve. Also notice that the second and third control points *do not lie on the curve*.

To see how the second and third control points affect the shape of the curve, let's add a pair of line segments which we'll call "handles." One handle connects the first and second control points, and the other handle connects the third and fourth control points.

```
handles = Graphics[{Red,
    Line[controlPts[[{1, 2}]]], Line[controlPts[[{3, 4}]]]}];
Show[dots, bezCurve, handles, Frame → True]
```

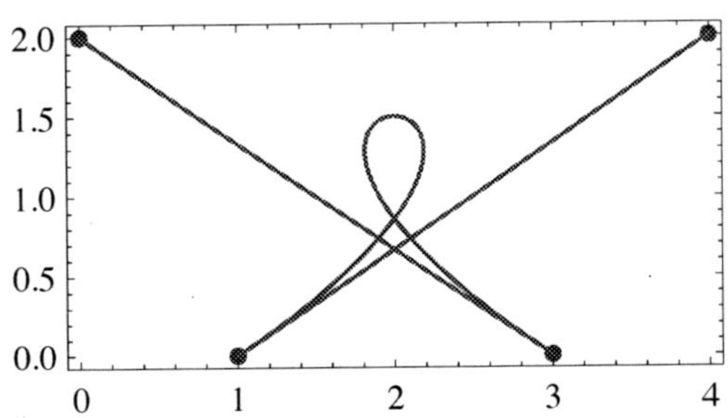

You should notice here that the handles are tangent to the curve at the first and fourth control points. This partially indicates the role of the second and third control points in determining the shape of the curve. It turns out that the length of a handle is important as well. (If you have used graphics applications such as Adobe Illustrator™, then you may already be familiar with such control points and handles.)

The following function combines the elements of the preceding construction and allows the specification of additional options. (The argument `opts` is followed by three "blanks.")

```
bezier[ctrlPts_, opts___] := Show[ParametricPlot[
   {(1 - t)^3, 3 (1 - t)^2 t, 3 (1 - t) t^2, t^3}.ctrlPts, {t, 0, 1}],
  Graphics[{{PointSize[.02], Red, Point /@ ctrlPts},
    {Red, Line[ctrlPts[[{1, 2}]]], Line[ctrlPts[[{3, 4}]]] } } ],
  opts, Frame → True, Axes → None, PlotRange → All]
```

Now the plot shown above can be created by simply entering

```
bezier[{{1, 0}, {4, 2}, {0, 2}, {3, 0}}]
```

We can also easily create an manipulator in which you can click and drag each of the control points to see how moving the control points affects the shape of the curve.

```
Manipulate[bezier[{p1, p2, p3, p4},
   Frame → None, PlotRange → {{-.1, 4.1}, {-.1, 3.1}}],
  {{p1, {1, 0}}, Locator}, {{p2, {0, 3}}, Locator},
  {{p3, {4, 3}}, Locator}, {{p4, {3, 0}}, Locator}]
```

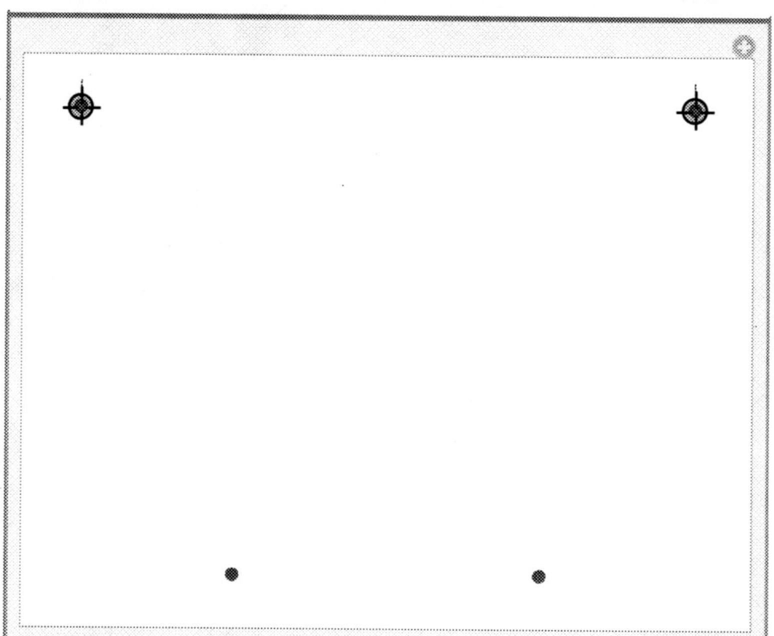

Arc Length

The length of a parametric curve

$$x = f(t),\ y = g(t),\ a \le t \le b,$$

is given by

$$L = \int_a^b \sqrt{(x'(t))^2 + (y'(t))^2}\, dt.$$

provided that the curve is traced out exactly once as t increases from a to b. (Otherwise, L may still be interpreted as total distance travelled between time $t = a$ and time $t = b$ by a particle whose position at time t is given by $(f(t), g(t))$.)

Example 7.1.4

Find the length of one arch of the cycloid in Example 7.1.1.

Let's first define functions and plot the curve.

```
x[t_] := t - Sin[t]; y[t_] := 1 - Cos[t]
ParametricPlot[{x[t], y[t]}, {t, 0, 2 π}]
```

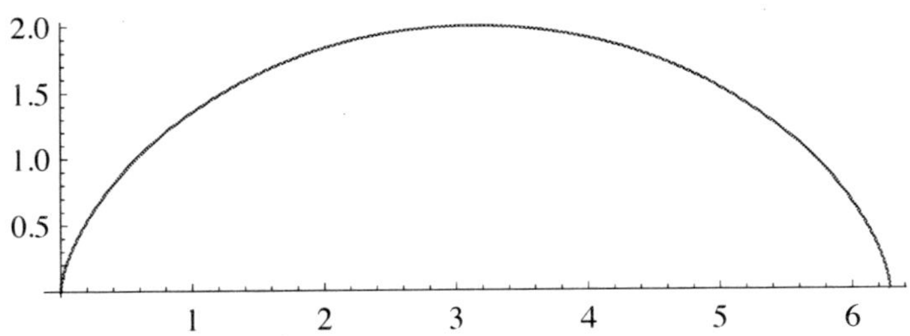

The computation of the integrand and subsequent integral is as follows.

```
√(x′[t]² + y′[t]²) // Simplify
∫₀^(2 π) % ⅆt
```

```
√(2 - 2 Cos[t])
8
```

So, oddly enough, the length of the curve is exactly 8.

■ Example 7.1.5

Find the length of the Lissajous figure given by $x = \cos t$, $y = \sin 3t$.

We proceed just as in the previous example.

```
x[t_] := Cos[t]; y[t_] := Sin[3 t]
ParametricPlot[{x[t], y[t]}, {t, 0, 2 π}]
```

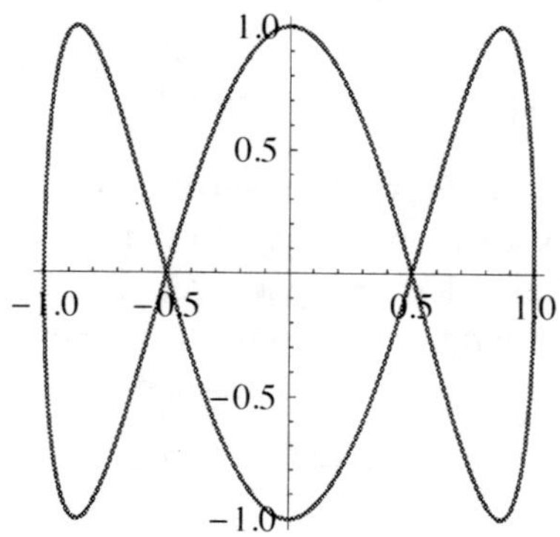

$$\sqrt{x'[t]^2 + y'[t]^2} \text{ // Simplify}$$

$$\int_0^{2\pi} \% \, dt$$

$$\sqrt{9 \cos[3t]^2 + \sin[t]^2}$$

$$\int_0^{2\pi} \sqrt{9 \cos[3t]^2 + \sin[t]^2} \, dt$$

So *Mathematica* is unable to evaluate the arc-length integral symbolically. A numerical value is computed as follows, where we use symmetry and integrate over a shorter interval.

$$4 \text{ NIntegrate}\left[\sqrt{x'[t]^2 + y'[t]^2}, \{t, 0, \pi/2\}\right]$$

```
13.0654
```

◆ Exercises

1. Plot and find the length of the curve given by

$$x = \cos t, \quad y = \sin t^2, \quad 0 \le t \le \sqrt{3\pi}.$$

2. Plot and find the length of the curve given by

$$x = 2\cos t, \quad y = \sin 2t.$$

3. A baseball player hits a long home-run in which the ball's path in flight is described by

$$x = 522\left(1 - e^{-t/3}\right), \quad y = 588\left(1 - e^{-t/3}\right) - 96t + 3$$

until the ball is caught by a fan at a point 30 feet above the ground. Find the time t at which the ball is caught, and find the distance that the ball travels in flight. Then compute the average speed of the ball during its flight.

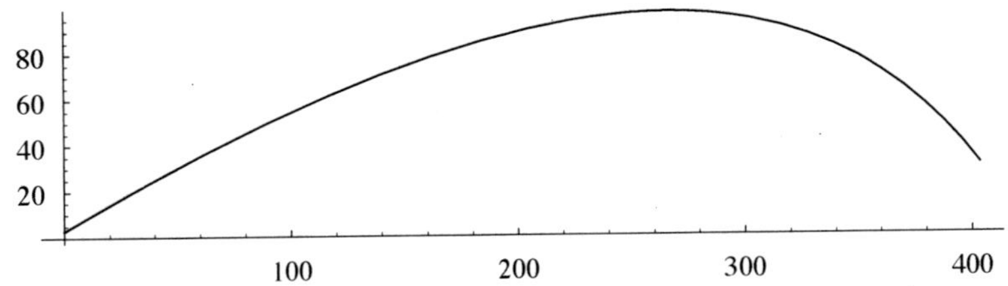

4. Compute the length of the first Bézier curve in Example 7.1.3.

5. The planet Fyodor has a circular orbit about its sun, and Fyodor's moon Theo has circular orbit about Fyodor in the same plane. Theo revolves about Fyodor ten times per Fyodoran year, and the distance from Theo to Fyodor is one-seventh the distance from Fyodor to its sun. Assuming that the sun is located at the origin, determine appropriate parametric equations and plot the path of Theo. (*Note*: Physics is totally ignored in this problem!)

6. Use the following code to produce an interesting animation of Lissajous figures.

```
Animate[
 ParametricPlot[{Cos[t], Sin[n t + k]}, {t, 0, 2 π}, Axes → None,
  PlotStyle → {Thick, Green}, Background → Black],
 {k, 0, 2 π}, {n, 1, 8, 1, ControlType → SetterBar}]
```

7.2 Polar Curves

A **polar curve** is the graph of an equation $r = f(\theta)$ where r and θ are standard polar coordinates defined by

$$x = r\cos\theta,\ y = r\sin\theta.$$

Mathematica's `PolarPlot` function plots polar curves.

■ Example 7.2.1

The graph of $r = \ln(\theta + 1)$ is a *logarithmic spiral*.

```
PolarPlot[Log[θ + 1], {θ, 0, 5 π}]
```

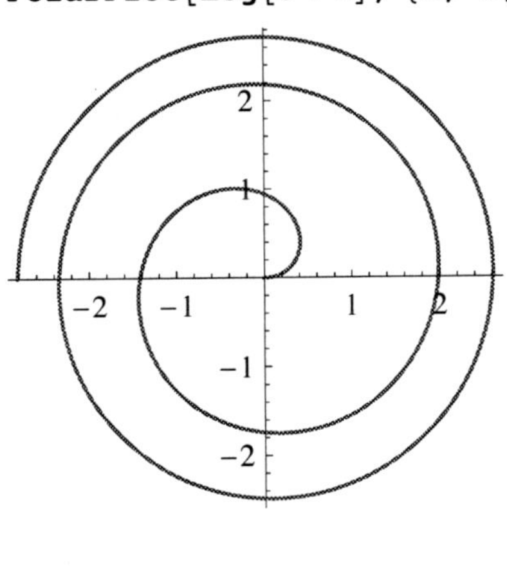

Example 7.2.2

The graph of $r = \sin 5\theta$ is a *five-leaved rose*. Note that the entire figure is drawn as θ goes from 0 to π. The curve would be trace out twice with $0 \le \theta \le 2\pi$. (Try it.)

```
PolarPlot[Sin[5 θ], {θ, 0, π}]
```

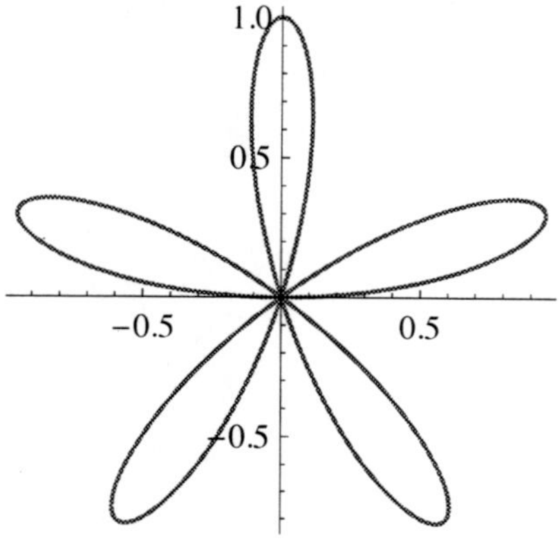

■ Animating a Polar Plot

In order to get a better understanding polar graphs, it is worth our while to animate the plotting of a couple of examples. We'll first make a function for plotting a *polar grid*:

```
polarGrid[r_, dr_] :=
 Graphics[{Gray, Table[Circle[{0, 0}, k], {k, dr, r, dr}], Table[
    Line[{{0, 0}, r {Cos[t], Sin[t]}}], {t, π / 12, 2 π, π / 12}]}]
```

For example,

```
polarGrid[1, .1]
```

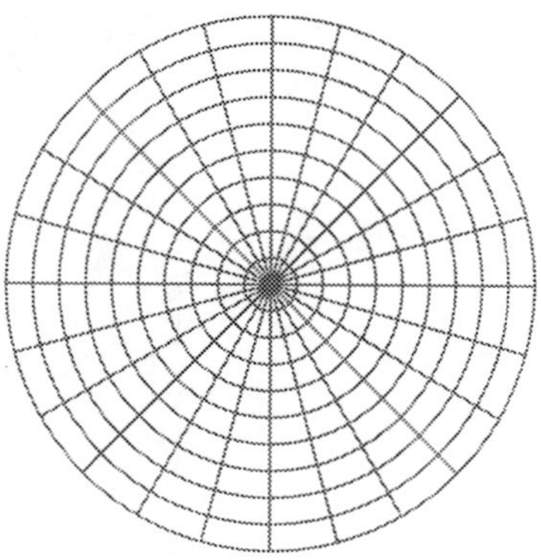

The following program creates a plot containing a polar grid and a polar curve $r = f(\theta)$ for $\theta_1 \le \theta \le \theta_2$, as well as the ray $\theta = \theta_2$ and a "radius" extending from the origin to the point (r_2, θ_2). (Be sure to enter `polarGrid` before using this program.)

```
polarCurve[r_, {θ_, θ1_, θ2_}, rmax_, opts___] :=
 Module[{r2 = r /. θ → θ2, pt2 = {Cos[θ2], Sin[θ2]}},
  Show[polarGrid[rmax, rmax / 10],
   PolarPlot[r, {θ, θ1, θ2}], Graphics[
    {{Purple, Line[{{0, 0}, rmax pt2}]}, {If[r2 > 0, Green, Red],
      Thick, Opacity[.5], Line[{{0, 0}, r2 pt2}]},
     {PointSize[Medium], Point[r2 pt2]}}],
   opts, PlotRange → 1.02 rmax {{-1, 1}, {-1, 1}}]]
```

For example, this plots the curve $r = \cos\theta$ for $0 \le \theta \le \pi/4$:

```
polarCurve[Cos[θ], {θ, 0, π / 3}, 1]
```

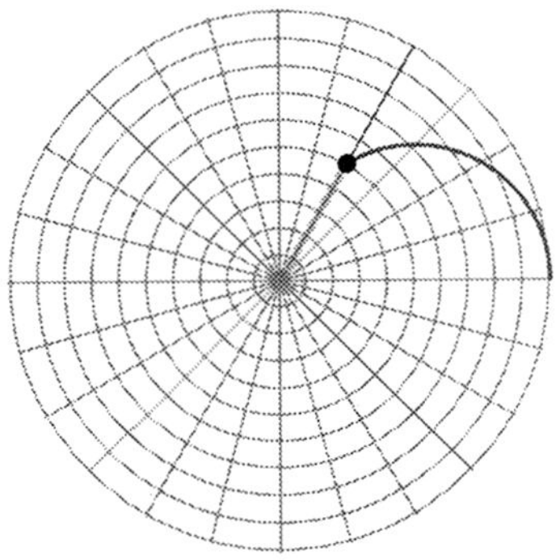

Now we're ready to animate. The following program uses `polarCurve` to create an animation of plots of a polar curves $r = f(\theta)$. The polar curve is plotted in a `GraphicsRow` alongside the corresponding rectangular plot of $y = f(x)$.

```
polarMovie[r_, {θ_, θ1_, θ2_}, rmax_] :=
 Animate[GraphicsRow[{polarCurve[r, {θ, θ1, k}, rmax],
    Plot[r, {θ, θ1, k}, Ticks → {{π, 2 π}, None},
     PlotRange → {{θ1, θ2}, {-rmax, rmax}}, ImageSize → 150]}, 20],
  {k, θ1 + π / 12, θ2, π / 12}]
```

This creates the animation for $r = \cos 2\theta, 0 \le \theta \le 2\pi$.

```
polarMovie[Cos[2 θ], {θ, 0, 2 π}, 1] // boldEdge
```

Arc Length

The length of a polar curve $r = f(\theta)$, where $\alpha \le \theta \le \beta$, is

$$L = \int_{\alpha}^{\beta} \sqrt{f(\theta)^2 + f'(\theta)^2}\ d\theta .$$

Example 7.2.3

Find the length of the four-leaved rose:

```
r[θ_] := Cos[2 θ];
Show[polarGrid[1.1, .1], PolarPlot[r[θ], {θ, 0, 2 π}], Axes → True]
```

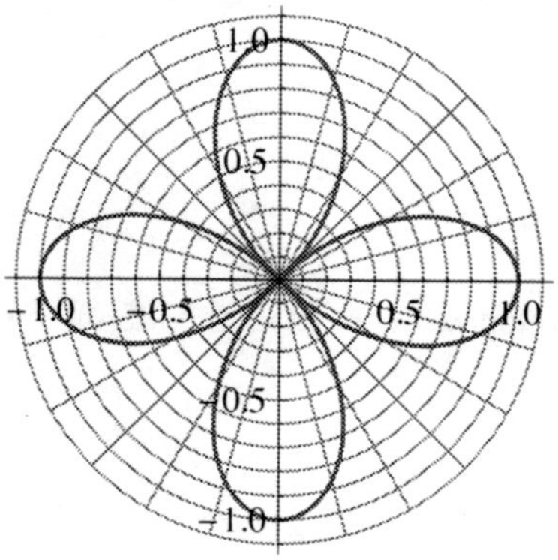

Because of symmetry, one-eighth of the length is traced out between $\theta = 0$ and $\theta = \frac{\pi}{4}$.

```
polarCurve[Cos[2 θ], {θ, 0, π / 4}, 1]
```

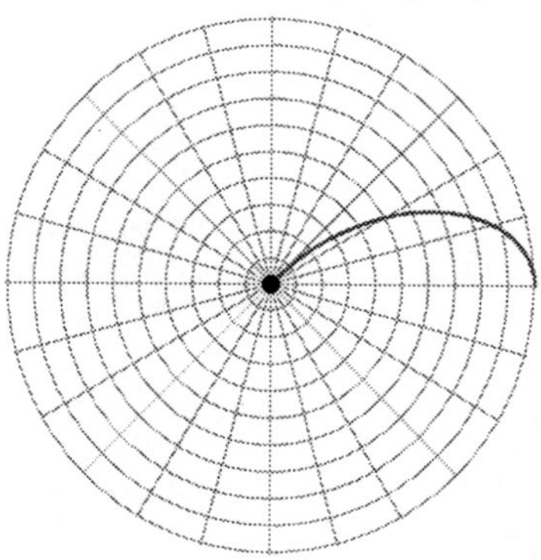

So let's integrate over $0 \le \theta \le \frac{\pi}{4}$ and multiply the result by eight. The integrand of interest and subsequent arc length computation are as follows.

```
√(r[θ]² + r'[θ]²)
{arclen = 8 ∫_0^(π/4) % dθ, arclen // N}
√(Cos[2 θ]² + 4 Sin[2 θ]²)
{4 EllipticE[-3], 9.68845}
```

■ Example 7.2.4

Find the length of the spiral $r = e^{-\theta/5}$, $0 \le \theta < \infty$.

```
r[θ_] := e^(-θ/5);
Show[polarGrid[1.1, .1], PolarPlot[r[θ], {θ, 0, 5 π}], Axes → True]
```

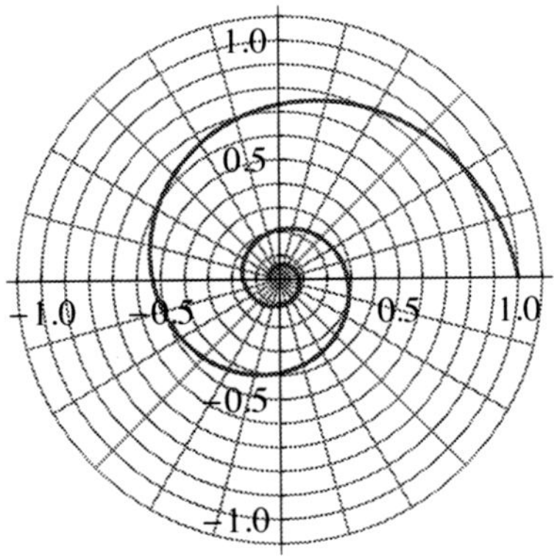

Note that we are dealing here with an improper integral. The integrand of interest and subsequent arc length computation are as follows.

$$\sqrt{r[\theta]^2 + r'[\theta]^2}$$

$$\left\{\text{arclen} = \int_0^\infty \% \, d\theta, \ \text{arclen} \, // \, N\right\}$$

$$\frac{1}{5}\sqrt{26}\sqrt{e^{-2\theta/5}}$$

$$\left\{\sqrt{26}, 5.09902\right\}$$

■ Plotting a "Polar Region"

Our goal here is to modify the `polarCurve` function above in order to plot and shade a region bounded by two polar curves. This will require a special graphic primitive for shading a "polar quadrilateral":

```
polarQuad[{α_, β_}, {r1α_, r1β_}, {r2α_, r2β_}] :=
  Polygon[{r1α {Cos[α], Sin[α]}, r2α {Cos[α], Sin[α]},
    r2β {Cos[β], Sin[β]}, r1β {Cos[β], Sin[β]}}]
```

For example,

```
Graphics[{LightGray, polarQuad[{π/6, π/4}, {.3, .4}, {1, .8}]}]
```

The following program does the job for us. (Again, be sure to enter `polarQuad` before using this program.)

```
polarRegion[{r1_, r2_}, {θ_, θ1_, θ2_}, rmax_, opts___] :=
 Module[{f, g, curves, rays, radius, shading, dθ = π / 36., p2},
  g[x_] := r1 /. θ → x; f[x_] := r2 /. θ → x; p2 = {Cos[θ2], Sin[θ2]};
  curves = PolarPlot[{r1, r2}, {θ, θ1, θ2}];
  rays =
   Graphics[{Blue, Line[rmax {{Cos[θ1], Sin[θ1]}, {0, 0}, p2}]}];
  radius = Graphics[{If[f[θ2] - g[θ2] > 0, Green, Red], Thick,
     Opacity[.5], Line[{g[θ2] p2, f[θ2] p2}]}]; shading =
   Table[Graphics[{Opacity[.1], Blue, polarQuad[{t - dθ, t},
       {g[t - dθ], g[t]}, {f[t - dθ], f[t]}]}], {t, θ1 + dθ, θ2, dθ}];
  Show[shading, polarGrid[rmax, rmax / 10], rays, curves,
   radius, opts, PlotRange → 1.02 rmax {{-1, 1}, {-1, 1}}]]
```

Now, for example, to create a plot that shows the region bounded by $r = \frac{2}{3}\cos\theta$, $r = \cos\theta$, $\theta = 0$, and $\theta = \pi/4$, we can enter this:

```
polarRegion[{2 Cos[θ] / 3, Cos[θ]}, {θ, 0, π / 4}, 1]
```

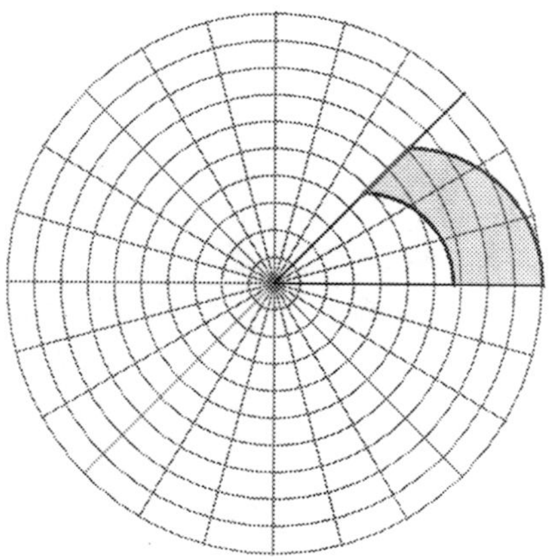

■ Area

The area of a region bounded by the graph of $r = f(\theta)$ and rays $\theta = \alpha$ and $\theta = \beta$ is given by

$$\int_{\alpha}^{\beta} \frac{1}{2} f(\theta)^2 \, d\theta,$$

provided that f is continuous and nonnegative and that $0 \leq \beta - \alpha \leq 2\pi$.

■ Example 7.2.5

Find the area enclosed by one loop of the three-leaved rose:

$$r = \sin 3\theta.$$

One loop of the figure is drawn as θ rotates from $\theta = 0$ to $\theta = \pi/3$, as is seen with the help of `polarRegion`:

```
polarRegion[{0, Sin[3 θ]}, {θ, 0, π / 3}, 1, Axes → True]
```

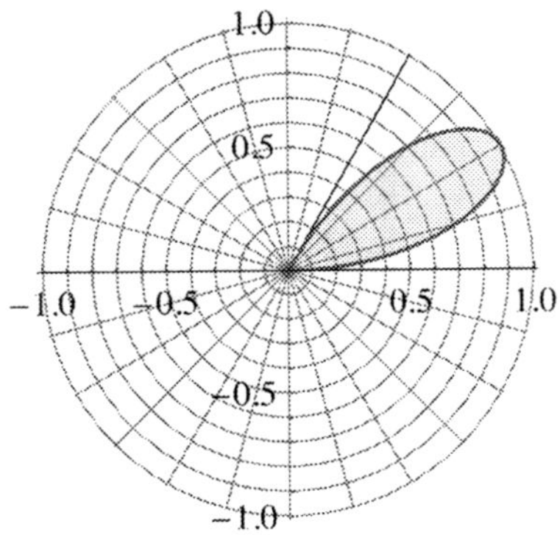

Therefore, the area of the enclosed region is

$$\int_0^{\pi/3} \frac{1}{2} \text{Sin}[3\,\theta]^2 \, d\theta$$

$$\frac{\pi}{12}$$

The area of a region bounded by two polar curves $r = f(\theta)$ and $r = g(\theta)$ and the rays $\theta = \alpha$ and $\theta = \beta$ is given by

$$\int_\alpha^\beta \frac{1}{2}\left(f(\theta)^2 - g(\theta)^2\right) d\theta,$$

assuming that f and g are continuous and nonnegative, $0 \le \beta - \alpha \le 2\pi$, and $f(\theta) \ge g(\theta)$ for $\alpha \le \theta \le \beta$.

■ Example 7.2.6

Find the area of the region that lies inside the circle $r = 3 \sin\theta$ and outside the circle $r = 2$.

Let's begin by shading the region between the curves for $0 \le \theta \le \pi$.

```
polarRegion[{2, 3 Sin[θ]}, {θ, 0, π}, 3]
```

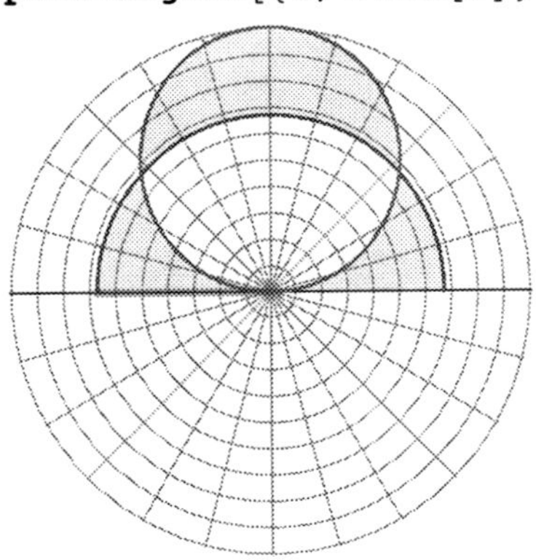

Next we need to find the angles that correspond to the points where the curves intersect. Setting $3\sin\theta = 2$, we can easily see that the curve intersect first when $\theta = \sin^{-1}(2/3)$ and again when $\theta = \pi - \sin^{-1}(2/3)$.

```
polarRegion[{2, 3 Sin[θ]}, {θ, ArcSin[2 / 3], π - ArcSin[2 / 3]}, 3]
```

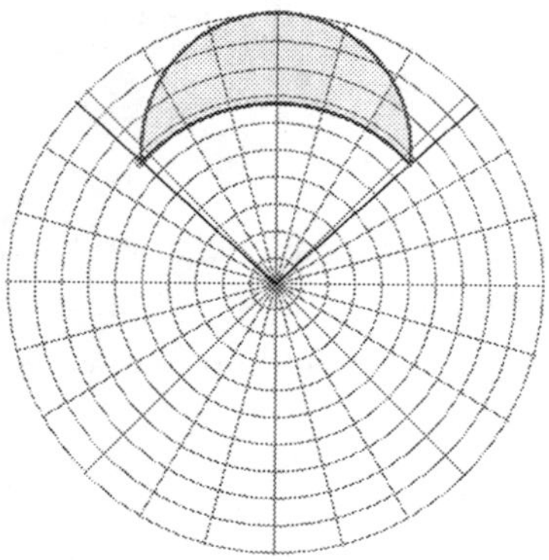

Taking advantage of the region's symmetry about the y-axis, we now compute its area:

```
area = 2 ∫_{ArcSin[2/3]}^{π/2} 1/2 (9 Sin[θ]^2 - 4) dθ;
area
area // N
```

$\frac{1}{2}\left(2\sqrt{5} + \text{ArcCos}\left[\frac{2}{3}\right]\right)$

2.6566

◆ Exercises

1. Use `polarMovie` to animate the polar curves in Examples 7.2.1, 7.2.2 and 7.2.4.
2. Use `polarMovie` to animate the *limaçon* $r = 1 + 2\cos\theta$.
3. Use `polarMovie` to animate the polar curve $r = \sin\theta\,\cos 2\theta$. Then compute its length and the area of the region it encloses.
4. Plot the *limaçon* $r = 3 + 2\cos\theta$ and compute its circumference.
5. Plot the polar curve $r = \theta(2-\theta)(3-\theta)$ and find the area inside each of its two loops.
6. Find the area inside one loop of the five-leaved rose $r = \sin 5\theta$.
7. Plot the region inside the *cardioid* $r = 1 - \cos\theta$ and compute its area.
8. Plot the region inside the inner loop of the *limaçon* $r = 1 + 2\cos\theta$ and compute its area.
9. Find the area of the region inside the circle $r = 3\cos\theta$ and outside the circle $r = 2\sin\theta$.
10. Use the defining polar relations $x = r\cos\theta$ and $y = r\sin\theta$ to plot the *nephroid of Freeth* $r = 1 + 2\sin(\theta/2)$, $0 \le \theta \le 4\pi$, using `ParametricPlot` (rather than `polarCurve` or `PolarPlot`).

8 Sequences and Series

Mathematica has graphical, algebraic, and numerical capabilities that lend themselves to the study of sequences and series. This chapter is an introduction to these capabilities.

8.1 Sequences

This section is devoted to the study of sequences. We begin with a look at *Mathematica*'s ability to generate sequences and compute limits symbolically.

In *Mathematica*, a sequence is simply a list. As we have seen before, `Table` often provides a convenient way to generate a list.

■ Example 8.1.1

Consider the sequence defined by

$$a_n = \frac{3n^2 - 2}{7n^2 - 10n + 5}, \quad n = 1, 2, 3, \ldots .$$

Such a sequence is really just the (ordered) range of the function $f(n)$ that defines a_n. The following defines that function and generates the first twenty terms of the sequence.

```
f[n_] := (3 n^2 - 2) / (7 n^2 - 10 n + 5)
seq = Table[f[n], {n, 1, 20}] // N
```

```
{0.5, 0.769231, 0.657895, 0.597403, 0.561538, 0.538071, 0.521583,
 0.509383, 0.5, 0.492562, 0.486523, 0.481523, 0.477316, 0.473727,
 0.470629, 0.467929, 0.465554, 0.46345, 0.461571, 0.459885}
```

A plot of the sequence can created with `ListPlot`.

```
ListPlot[seq, PlotRange → {0, 1}]
```

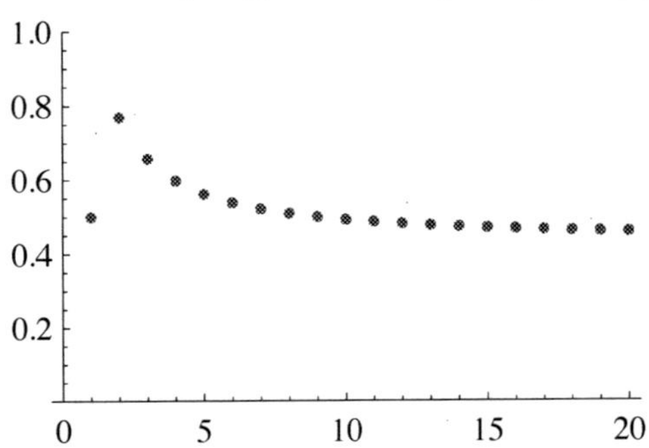

The limit of the sequence (as $n \to \infty$) can be found by applying `Limit` to the function f.

```
Limit[f[n], n → ∞]
```

$$\frac{3}{7}$$

Example 8.1.2

Consider the sequence defined by

$$a_n = \frac{3n + 7\,(-1)^n}{4n + 5}, \quad n = 1, 2, 3, \ldots .$$

The following defines the associated function and generates the first twenty-five terms of the sequence.

```
f[n_] := (3 n + 7 (-1)^n) / (4 n + 5)
seq = Table[f[n], {n, 1, 40}] // N
```

```
{-0.444444, 1., 0.117647, 0.904762, 0.32, 0.862069, 0.424242, 0.837838, 0.487805,
 0.822222, 0.530612, 0.811321, 0.561404, 0.803279, 0.584615, 0.797101, 0.60274,
 0.792208, 0.617284, 0.788235, 0.629213, 0.784946, 0.639175, 0.782178,
 0.647619, 0.779817, 0.654867, 0.777778, 0.661157, 0.776, 0.666667, 0.774436,
 0.671533, 0.77305, 0.675862, 0.771812, 0.679739, 0.770701, 0.68323, 0.769697}
```

A plot of the sequence can created with `ListPlot`.

```
seqPlot = ListPlot[seq, PlotRange → {0, 1.1}]
```

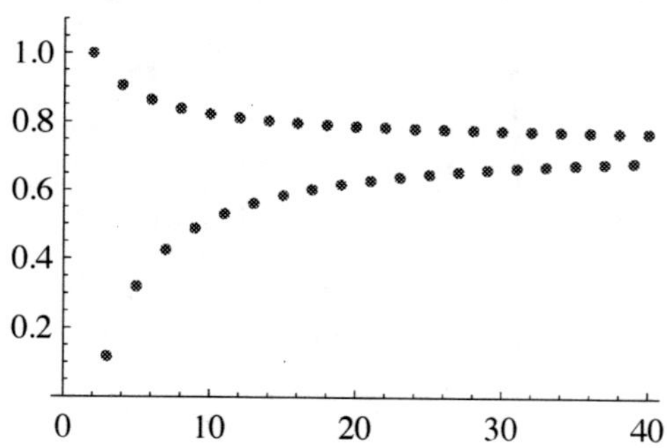

This time `Limit` is not able to find the limit of the sequence.

```
Limit[f[n], n → ∞]
```

$\frac{3}{4}$

But if we look much farther out in sequence, we see

```
seq = Table[f[n], {n, 10000, 10010}] // N
```

```
{0.750081, 0.749731, 0.750081, 0.749731, 0.750081,
 0.749731, 0.750081, 0.749731, 0.750081, 0.749732, 0.750081}
```

and it becomes evident that the limit of the sequence is $3/4$. This limit can be proven by means of the *Squeeze Theorem*, but it essentially follows from the fact that for larger and larger n, the $7\,(-1)^n$ term in the numerator and the 5 in the denominator have less and less effect upon the value of $f(n)$; thus for large n,

$$f(n) \approx \frac{3n}{4n} = \frac{3}{4},$$

and the approximation improves as n increases.

Recursive Sequences

In the preceding examples, the nth term of the sequence depends only upon n. Sequences in which the nth term depends not upon n but upon previous terms in the sequence are far more interesting. These are called **recursive sequences**. *Mathematica*'s tool for computing terms of a recursive sequence is `NestList`.

Example 8.1.3

The sequence defined by

$$\begin{aligned} a_1 &= .5 \\ a_{n+1} &= 2\sin a_n, \quad n = 1, 2, \ldots \end{aligned}$$

is a recursive sequence. To compute twenty terms of this sequence with `NestList`, we first define the "generating function" of the sequence,

```
g[x_] := 2 Sin[x]
```

and then enter

```
NestList[g, .5, 25]

{0.5, 0.958851, 1.63706, 1.99561, 1.82223, 1.93711, 1.86731,
 1.91272, 1.88422, 1.90257, 1.89093, 1.89838, 1.89364,
 1.89667, 1.89474, 1.89597, 1.89519, 1.89569, 1.89537,
 1.89557, 1.89544, 1.89553, 1.89547, 1.89551, 1.89549, 1.8955}
```

From these numbers, we observe that the sequence seems to converge to a limit approximately equal to 1.8955. Notice that when we compute $g(1.8955)$ and round to decimal 5 places, we see

```
N[g[1.8955], 5]

1.89549
```

So it looks as though the limit of the sequence satisfies the equation $x = g(x)$.

Just as we observed in the preceding example, if a recursive sequence generated by g,

$$a_{n+1} = g(a_n), \quad n = 1, 2, \ldots,$$

converges to a limit x^* (and if g is continuous), it follows that

$$x^* = g(x^*);$$

that is, x^* is a **fixed point** of g.

The generation of sequences such as this, in which $a_{n+1} = g(a_n)$ with some given continuous function g and some given value of a_1, is called *functional iteration*, or *fixed-point iteration*. There is a special graphical device, called a **web plot**, for visualizing a sequence generated in this way. A web plot consists essentially of the graphs of $y = x$ and $y = g(x)$ and alternating vertical and horizontal line segments joining the points

$$(a_1, a_1), (a_1, a_2), (a_2, a_2), (a_2, a_3), (a_3, a_3), (a_3, a_4), (a_4, a_4), \ldots,$$

which fall alternately on the graphs of $y = x$ and $y = g(x)$.

The following program draws web plots.

```
webPlot[g_, {x_, xmin_, xmax_}, {a1_, n_}, opts___Rule] :=
 Module[{seq, pts}, seq = NestList[(g /. x → #) &, N[a1], n];
  pts = Join[{{a1, 0}}, Flatten[Partition[seq, 2, 1] /.
      {s_?NumericQ, t_} → {{s, t}, {t, t}}, 1]];
  Show[Graphics[{{Red, Line[pts]}, Text[ToString[Last[seq]],
      {(xmin + xmax) / 2, .1 xmin + .9 xmax}]}],
   Plot[{x, g}, {x, xmin, xmax}, opts],
   FilterRules[{opts}, Options[Graphics]], Frame → True,
   PlotRange → {{xmin, xmax}, {xmin, xmax}}]]
```

Example 8.1.4

Consider the sequence

$$a_1 = 0.2$$
$$a_{n+1} = \cos a_n, \quad n = 1, 2, \ldots$$

The following is the associated web plot, constructed from the first seventeen terms of the sequence.

```
cosweb = webPlot[Cos[x], {x, 0, 1.2}, {.2, 16}]
```

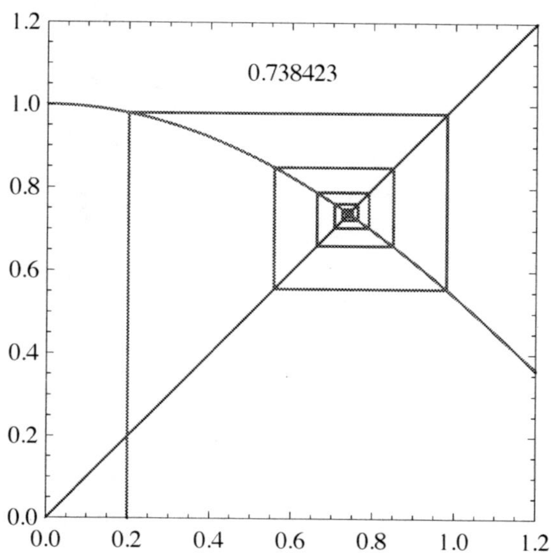

It is interesting to view the web plot alongside a standard sequence plot. Note that the vertical axes are identical. Consequently the horizontal segments in the web plot line up with corresponding points in the sequence plot.

```
GraphicsGrid[{{cosweb, ListPlot[NestList[Cos, .2, 16],
   PlotRange → {0, 1.2}, AspectRatio → 1]}}]
```

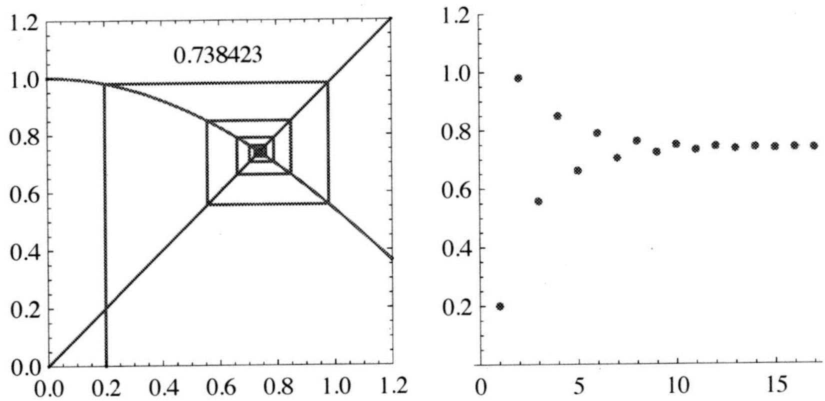

Moving the slider left to right here constructs the web plot one iteration at a time:

```
Manipulate[webPlot[Cos[x], {x, 0, 1.2}, {.2, k}], {k, 1, 12, 1}]
```

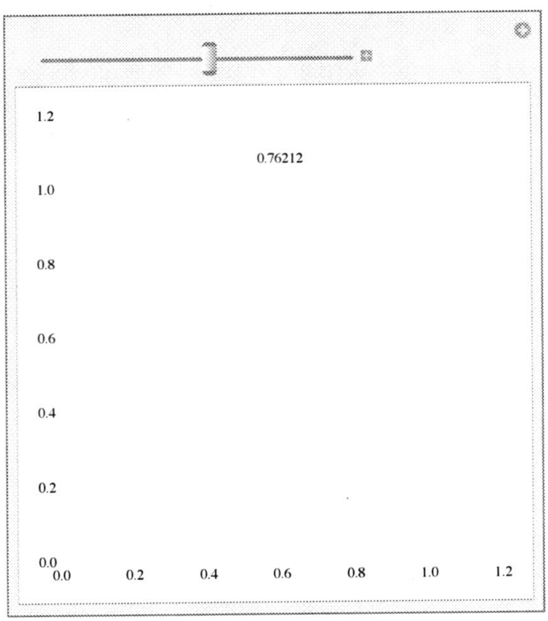

The web plot indicates that the sequence converges to a solution of the equation $x = \cos x$. By examining the values of $a_{10}, a_{25}, a_{50}, a_{75}, a_{100}$ (for example),

```
NestList[Cos, .2, 100][[{10, 25, 50, 75, 100}]]

{0.749577, 0.739057, 0.739085, 0.739085, 0.739085}
```

we observe that the limit of the sequence—and thus the solution of $x = \cos x$—is 0.739085 when rounded to six decimal places. This is confirmed by `FindRoot`:

```
FindRoot[x - Cos[x], {x, 1}]
```

```
{x → 0.739085}
```

This is an example of a very important use of sequences. Exact solutions of many equations simply cannot be found. Thus we resort to an approximation procedure that amounts to generating a sequence that (we hope) converges to a solution. Provided that the sequence does converge, by computing enough terms in the sequence we can approximate the exact solution to any desired accuracy.

▪ Example 8.1.5

Consider the sequence

$$a_1 = 2.5$$
$$a_{n+1} = a_n(2.75 - a_n),\ n = 1, 2, \ldots$$

The following is the associated web plot, constructed from the first twenty terms of the sequence.

```
webPlot[x (2.75 - x), {x, 0, 2.75}, {2.5, 19}]
```

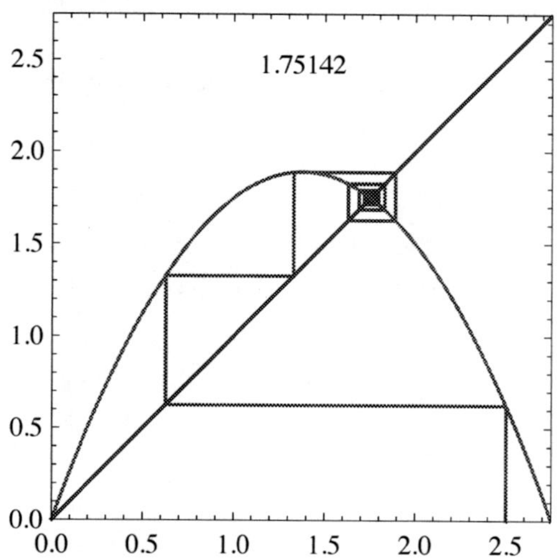

So clearly the sequences converges. Its limit satisfies the quadratic equation $x = x(2.75 - x)$, and thus is exactly 1.75.

▪ Example 8.1.6

Consider the sequence

$$a_1 = 1.5$$
$$a_{n+1} = a_n(4 - a_n),\ n = 1, 2, \ldots$$

The following is the associated web plot, constructed from the first 100 terms.

```
webPlot[x (4 - x), {x, 0, 4.75}, {1.5, 99}]
```

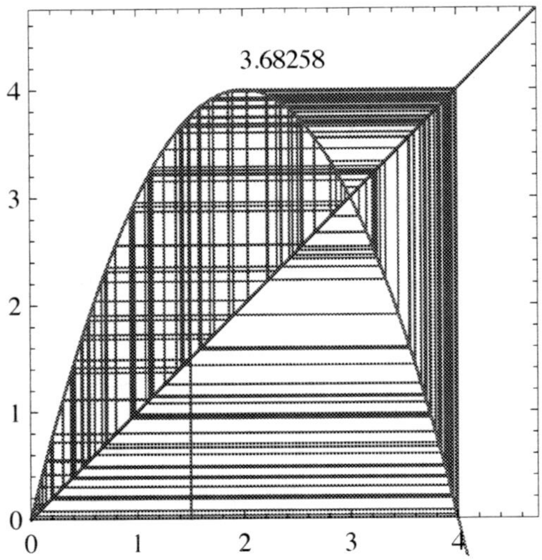

Clearly this sequence does not converge. Indeed, its terms jump about quite chaotically between 0 and 4.

Newton's Method Revisited

Recall that Newton's method for approximating a solution of $f(x) = 0$ is described by

$$x_{n+1} = x_n - \frac{f(x_k)}{f'(x_k)}, \quad k = 0, 1, 2, \ldots$$

where x_0 is some chosen initial approximation to the solution. This procedure generates a *recursive sequence*, since $x_{n+1} = g(x_n)$, where the generating function g is defined by

$$g(x) = x - \frac{f(x)}{f'(x)}.$$

Note that if $x_n \to x^*$ as $n \to \infty$ and if $f'(x^*) \neq 0$, then

$$x^* = x^* - \frac{f(x^*)}{f'(x^*)},$$

which implies that $f(x^*) = 0$; i.e., x^* is a root of f.

Example 8.1.7

Consider the problem of approximating the solution of

$$x^3 + x - 1 = 0.$$

Examination of the graph of the function

```
f[x_] := x^3 + x - 1
```

shows that there is exactly one solution and that the solution is between .5 and 1.

```
Plot[f[x], {x, -1, 2}]
```

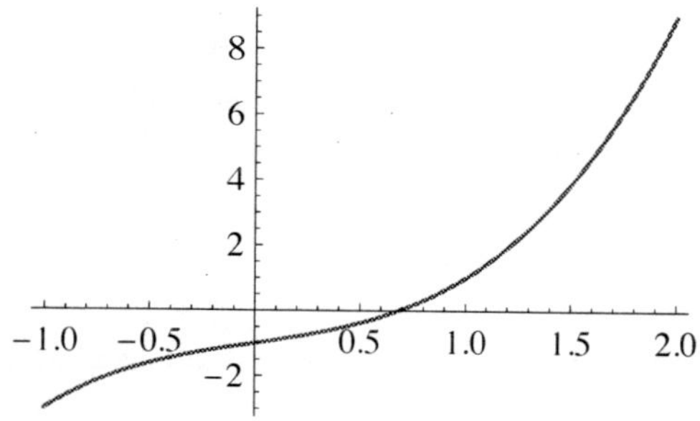

So let's take $x_0 = 1$ and use Newton's Method, which here takes the form

$$x_{n+1} = g(x_n), \quad n = 0, 1, 2, \ldots$$

where

```
g[x_] = x - f[x]/f'[x]
```

$$x - \frac{-1 + x + x^3}{1 + 3x^2}$$

The first six terms of the Newton sequence indicate its very rapid convergence:

```
NestList[g, 1., 5]

{1., 0.75, 0.686047, 0.68234, 0.682328, 0.682328}
```

Now let's look at the corresponding web plot.

```
webPlot[g[x], {x, 0, 1.5}, {1, 4}]
```

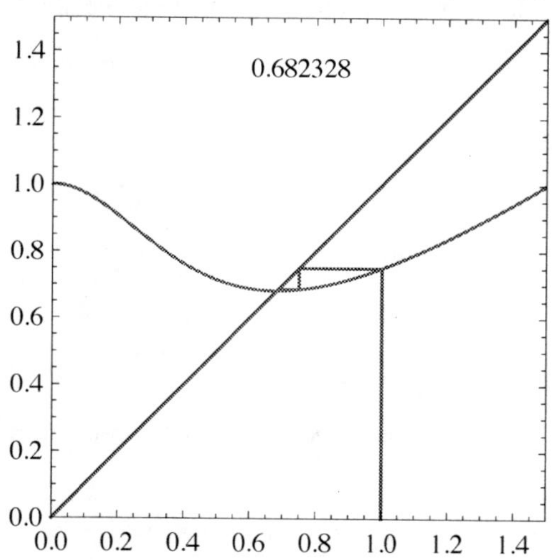

The curve in this web plot is the graph of the function

$$g(x) = x - \frac{f(x)}{f'(x)}.$$

Notice that the extremely rapid convergence of the sequence is due to the fact that the graph of g has a horizontal tangent at the point of intersection. One of the exercises that follow will ask you to verify that this is a general property of Newton's method.

◆ Exercises

For Exercises 1-6, plot each sequence and try to determine whether the sequence has a limit as $n \to \infty$ and, if so, estimate the value of the limit or make a conjecture about its exact value. Then have *Mathematica* (attempt to) compute the limit with `Limit`.

1. $a_n = \dfrac{2+(-1)^n n^2}{n^2 - 3n + 4}$
2. $a_n = \dfrac{\ln n}{n^{1/3}}$
3. $a_n = \dfrac{n!}{100^n}$
4. $a_n = \dfrac{2n^3 - 6n^2 + 15}{n^4 + 18n^3 - 6}$
5. $a_n = \left(\cos^2 3n\right)^{1/n}$
6. $a_n = \dfrac{n^4}{2^n}$

In Exercises 7-11, make a web plot of each recursive sequence and try to determine whether the sequence has a limit as $n \to \infty$ and, if so, estimate the value of the limit or make a conjecture about its exact value. **When possible, compute the *exact* value of the limit.** In all cases, describe as well as you can the long-term behavior of the sequence.

7. $a_{n+1} = 1 + \dfrac{a_n}{3},\ a_1 = 0$
8. $a_{n+1} = 1 + \dfrac{a_n}{3},\ a_1 = 3$
9. $a_{n+1} = 2a_n - 3,\ a_1 = 2.99$
10. $a_{n+1} = e^{-a_n},\ a_1 = 1$
11. $a_{n+1} = a_n(3 - a_n),\ a_1 = 1.5$

In Exercises 12-16, use Newton's method to approximate the smallest positive solution of the given equation to at least six decimal places.

12. $\sin^2 x - \cos x^2 = 0$
13. $x - \cos x = 0$
14. $x = e^{-x}$
15. $\tan x = x$
16. $x^5 - x^4 = 1$
17. Let $g(x) = x - f(x)/f'(x)$, where f is a given twice-differentiable function.

    a) Show that $g'(x) = \dfrac{f(x)\, f''(x)}{f'(x)^2}$.

    b) Conclude that if $f(x^*) = 0$ and $f'(x^*) \neq 0$, then $g'(x^*) = 0$.

    c) Describe the effect of $g'(x^*) = 0$ on the web plot of the Newton's method sequence.
18. Find the limit of this sequence of sums:

$$\frac{1}{2},\ \frac{1}{2}+\frac{4}{4},\ \frac{1}{2}+\frac{4}{4}+\frac{9}{8},\ \frac{1}{2}+\frac{4}{4}+\frac{9}{8}+\frac{16}{16},\ \frac{1}{2}+\frac{4}{4}+\frac{9}{8}+\frac{16}{16}+\frac{25}{32},\ \ldots$$

8.2 Series

For any sequence $\{a_n\}_{n=1}^{\infty}$, the corresponding **sequence of partial sums** $\{s_n\}_{n=1}^{\infty}$ is defined by

$$s_n = \sum_{k=1}^{n} a_k.$$

■ Example 8.2.1

If $a_n = 1/n^3$, then the corresponding sequence of partial sums is

$$\begin{aligned} s_1 &= 1, \\ s_2 &= 1 + \frac{1}{8} = \frac{9}{8}, \\ s_3 &= 1 + \frac{1}{8} + \frac{1}{27} = \frac{251}{216}, \\ s_4 &= 1 + \frac{1}{8} + \frac{1}{27} + \frac{1}{64} = \frac{2035}{1728}, \\ &\vdots \end{aligned}$$

The first ten partial sums are computed by *Mathematica* as follows.

$$\texttt{Table}\left[\sum_{k=1}^{n} \frac{1}{k^3}, \{n, 1, 20\}\right] \texttt{ // N}$$

```
{1., 1.125, 1.16204, 1.17766, 1.18566, 1.19029, 1.19321,
 1.19516, 1.19653, 1.19753, 1.19828, 1.19886, 1.19932,
 1.19968, 1.19998, 1.20022, 1.20043, 1.2006, 1.20074, 1.20087}
```

These numbers suggest that the sequence of partial sums converges to a limit approximately equal to 1.20. The 100th to the 110th partial sums are

$$\texttt{Table}\left[\sum_{k=1}^{n} \frac{1}{k^3}, \{n, 100, 110\}\right] \texttt{ // N}$$

```
{1.20201, 1.20201, 1.20201, 1.20201, 1.20201,
 1.20201, 1.20201, 1.20201, 1.20201, 1.20202, 1.20202}
```

These numbers suggest that the sequence of partial sums converges to a limit approximately equal to 1.202. (We still should withhold judgement on the next digit.) The 1000th and 2000th partial sums are

$$\left\{\sum_{k=1}^{1000} \frac{1.}{k^3}, \sum_{k=1}^{2000} \frac{1.}{k^3}\right\}$$

```
{1.20206, 1.20206}
```

and together suggest that the limit, rounded to six figures, is 1.20206.

Summing an Infinite Series

Given a sequence $\{a_n\}_{n=1}^{\infty}$, the formal expression

$$\sum_{k=1}^{\infty} a_k = a_1 + a_2 + a_3 + \cdots$$

is called an (infinite) series. The sum of an infinite series is the limit of the sequence of partial sums of $\{a_n\}_{n=1}^{\infty}$, provided that limit exists. In short, we define the sum of the series as

$$\sum_{k=1}^{\infty} a_k = \lim_{n\to\infty} \sum_{k=1}^{n} a_k .$$

For example, our investigations in the preceding example indicate that

$$\sum_{k=1}^{\infty} 1/k^3 \approx 1.20206 .$$

Mathematica is capable of computing the exact sum of many series. In fact, the sum of the series in Example 8.2.1 is given terms of a special function known as the *zeta function*:

$$\sum_{k=1}^{\infty} \frac{1}{k^3}$$

```
Zeta[3]
```

```
Zeta[3] // N
```

```
1.20206
```

Example 8.2.2

Each of the following three groups of computations shows the first ten summands in a series (i.e., the summands of the tenth partial sum), the first through tenth partial sums, and the exact sum of the series (i.e., the limit of the sequence of partial sums).

(i) $\sum_{k=1}^{\infty} \frac{1}{k^2}$

```
Table[1/k^2, {k, 1, 10}]
```

$$\left\{1, \frac{1}{4}, \frac{1}{9}, \frac{1}{16}, \frac{1}{25}, \frac{1}{36}, \frac{1}{49}, \frac{1}{64}, \frac{1}{81}, \frac{1}{100}\right\}$$

$$\text{Table}\left[\sum_{k=1}^{n} \frac{1.}{k^2}, \{n, 1, 10\}\right]$$

```
{1., 1.25, 1.36111, 1.42361, 1.46361,
 1.49139, 1.5118, 1.52742, 1.53977, 1.54977}
```

$$\sum_{k=1}^{\infty} \frac{1}{k^2}$$

$$\frac{\pi^2}{6}$$

(ii) $\sum_{k=1}^{\infty} \frac{1}{3^k}$

```
Table[3^-k, {k, 1, 10}]
```

$\left\{\frac{1}{3}, \frac{1}{9}, \frac{1}{27}, \frac{1}{81}, \frac{1}{243}, \frac{1}{729}, \frac{1}{2187}, \frac{1}{6561}, \frac{1}{19683}, \frac{1}{59049}\right\}$

$\text{Table}\left[\sum_{k=0}^{n} 3^{-k}, \{n, 0, 10\}\right] // N$

```
{1., 1.33333, 1.44444, 1.48148, 1.49383,
 1.49794, 1.49931, 1.49977, 1.49992, 1.49997, 1.49999}
```

$\sum_{k=0}^{\infty} 3^{-k}$

$\frac{3}{2}$

(iii) $\sum_{k=1}^{\infty} \frac{k^2}{2^k}$

```
Table[k^2 / 2^k, {k, 1, 10}]
```

$\left\{\frac{1}{2}, 1, \frac{9}{8}, 1, \frac{25}{32}, \frac{9}{16}, \frac{49}{128}, \frac{1}{4}, \frac{81}{512}, \frac{25}{256}\right\}$

$\text{Table}\left[\sum_{k=1}^{n} k^2 / 2^k, \{n, 1, 10\}\right] // N$

```
{0.5, 1.5, 2.625, 3.625, 4.40625,
 4.96875, 5.35156, 5.60156, 5.75977, 5.85742}
```

$\sum_{k=1}^{\infty} k^2 / 2^k$

6

▪ Example 8.2.3: Geometric Series

A **geometric series** is a series of the form

$$\sum_{k=0}^{\infty} r^k,$$

where $r > 0$. Such a series has a finite sum (i.e., the series *converges*) if and only if $-1 < r < 1$. Note however that if we enter

$\sum_{k=0}^{\infty} r^k$

$\frac{1}{1 - r}$

Mathematica returns a result based on the unstated *assumption* that $-1 < r < 1$. So the naïve user might conclude that

$$\sum_{k=0}^{\infty} 2^k = -1,$$

which is obviously nonsense.

■ Example 8.2.4: *p*-Series

A ***p*-series** is a series of the form

$$\sum_{k=1}^{\infty} \frac{1}{k^p},$$

where $p > 0$. Such a series has a finite sum (i.e., the series *converges*) if and only if $p > 1$. When $p = 1$, this is known as the **harmonic** series:

$$\sum_{k=1}^{\infty} \frac{1}{k}$$

Sum::div : Sum does not converge. ≫

$$\sum_{k=1}^{\infty} \frac{1}{k}$$

When p is only slightly larger than 1, the sum of the series is very large, but nevertheless the series converges. Consider the case where $p = 1.001$. By summing the first 1000 terms, we might expect to get a reasonable estimate of the sum.

$$\sum_{k=1}^{1000} \frac{1}{k^{1.001}}$$

```
7.46174
```

Let's compare that with the sum of 100,000 terms:

$$\sum_{k=1}^{100\,000} \frac{1}{k^{1.001}}$$

```
12.0242
```

What about 10^{10} terms?

$$\sum_{k=1}^{10^{10}} \frac{1}{k^{1.001}}$$

```
23.3401
```

It is pretty clear that we're getting nowhere fast, so let's just compute the following sums of 10^{1000}, 10^{2000}, 10^{3000}, …, $10^{10,000}$ terms.

$$\text{Table}\left[\sum_{k=1}^{10^{1000\,j}} \frac{1}{k^{1.001}},\ \{j,\ 1,\ 10\}\right]$$

```
{900.577, 990.577, 999.577, 1000.48, 1000.57,
 1000.58, 1000.58, 1000.58, 1000.58, 1000.58}
```

So at last it appears that with 10^{5000} terms we finally have a good estimate of the limit.

(Do you think it's possible for *Mathematica*—or any computer program—to literally add 10^{5000} numbers in a reasonable length of time? How many additions would have to be done per second to complete the job in one year? For information on how such a sum can be computed quickly, see the documentation for `NSum`.)

■ Convergence

It is often difficult to determine numerically whether a sequence converges. As we saw in Example 8.2.4, it is very difficult to determine by numerical investigations that the harmonic series diverges and that a p-series with p close to 1 converges. For this reason, we need analytical tests to determine whether a series converges. Moreover, when dealing with a convergent series, it is very important to have an estimate of the error when estimating the series with a partial sum.

There are five primary tests for convergence. These are the integral test, the comparison test, the limit-comparison test, the ratio test, and the root test. Because it provides a nice error estimate, we will illustrate only the integral test.

The Integral Test. Suppose that $a_n = f(n)$, where f is a positive, continuous, and decreasing function on the interval $[1, \infty)$. Then

$$\sum_{k=1}^{n} a_k \text{ converges if and only if } \int_1^{\infty} f(x)\,dx \text{ converges.}$$

Moreover, we have the error estimate

$$\int_{n+1}^{\infty} f(x)\,dx \leq \varepsilon_n \leq \int_n^{\infty} f(x)\,dx,$$

where $\varepsilon_n = \sum_{k=n+1}^{\infty} a_k$ is the difference between the sum of the series and the nth partial sum.

■ Example 8.2.5

Consider the series

$$\sum_{k=1}^{\infty} \frac{1}{n \ln n}.$$

According to the integral test, this series converges if and only if the improper integral $\int_1^{\infty} \frac{1}{x \ln x}\,dx$ converges. However, this integral does not converge, since the antiderivative of $(x \ln x)^{-1}$ has no limit as $x \to \infty$:

```
antiD = ∫ 1/(x Log[x]) dx
Log[Log[x]]
Limit[antiD, x → ∞]
∞
```

Thus the series in question is divergent.

■ Example 8.2.6

Consider the series

$$\sum_{k=1}^{\infty} \frac{n}{2^n}.$$

According to the integral test, the convergence or divergence of this series is determined by that of the improper integral $\int_1^{\infty} x\,2^{-x}\,dx$.

```
∫_1^∞ x 2^-x ⅆx
```

$$\frac{1 + \text{Log}[2]}{2\,\text{Log}[2]^2}$$

Thus the series does converge. Now suppose that we wish to estimate its sum to within 0.0005. How many terms do we need to sum in order to obtain that accuracy? Let's first compute $\int_n^{\infty} x\,2^{-x}\,dx$.

```
err[n_] = Assuming[n > 0, ∫_n^∞ x 2^-x ⅆx]
```

$$\frac{2^{-n}\,(1 + n\,\text{Log}[2])}{\text{Log}[2]^2}$$

We can now graphically determine the least value of n for which this error estimate is less than .0005.

```
Plot[{err[n], .0005}, {n, 10, 25}, PlotRange → {0, .001}]
```

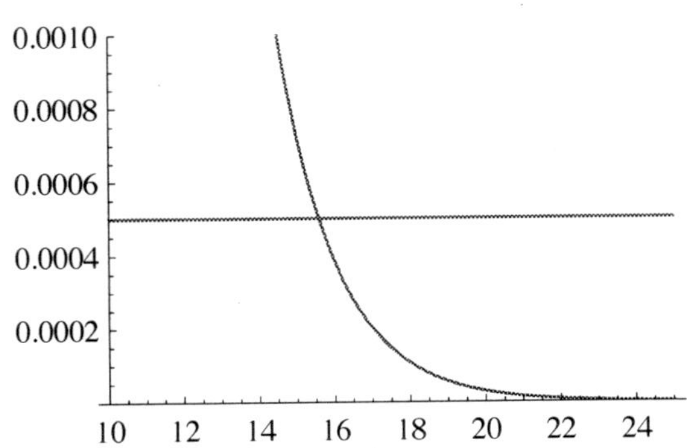

So the sum of sixteen terms will give us the desired accuracy.

```
Σ_{k=1}^{16} k 2^-k // N
```

1.99973

◆ Exercises

For each series in 1-6, plot the sequence of partial sums and estimate the sum of the series.

1. $\sum_{k=1}^{\infty} \frac{1}{k^2}$
2. $\sum_{k=1}^{\infty} \frac{k^2}{2^k}$
3. $\sum_{k=1}^{\infty} \frac{1}{k^k}$
4. $\sum_{k=0}^{\infty} \frac{(-1)^k}{(2k)!}\left(\frac{\pi}{3}\right)^{2k}$
5. $\sum_{k=0}^{\infty} \frac{(-1)^k}{(2k)!}\left(\frac{\pi}{2}\right)^{2k}$
6. $\sum_{k=0}^{\infty} \frac{1}{k!}$

For each series in 7-9, use the integral-test error estimate to determine how many terms, when summed, will estimate the series to within 0.0005.

7. $\sum_{k=1}^{\infty} \frac{\ln k}{k^2}$
8. $\sum_{k=1}^{\infty} \frac{\ln(1+k^2)}{k^2}$
9. $\sum_{k=1}^{\infty} \frac{1}{k(k+1)}$

8.3 Power Series

A power series, about the number a, is a *function* f whose value at x is the sum of a series:

$$f(x) = \sum_{k=0}^{\infty} c_k(x-a)^k = c_0 + c_1(x-a) + c_2(x-a)^2 + \cdots$$

for some given sequence of coefficients $c_0, c_1, c_2, \ldots$ Note that when $x = a$, the series consists of just one nonzero term and thus is guaranteed to converge. When $x \neq a$, the series may or may not converge, depending on the particular coefficients. So the domain of domain of such a function is the set of all x for which the series converges.

There are three possibilities for the domain. It is either:

(i) $\{a\}$, (ii) $(-\infty, \infty)$, or (iii) some bounded interval centered at a.

In cases (ii) and (iii), the domain is called the *interval of convergence* of the power series. In case (iii), the domain may be an open or closed interval, or it may contain one of its two endpoints. Half the length of this interval is called the radius of convergence. In case (i), the radius of convergence is zero, while in case (ii) the radius of convergence is ∞.

■ Example 8.3.1

Consider the power series

$$f(x) = \sum_{k=0}^{\infty} \frac{x^k}{k}.$$

The ratio test shows that the series converges (absolutely) if $-1 < x < 1$. The series diverges when $x < -1$ or $x \geq 1$ and is a convergent alternating series when $x = -1$. Thus the interval of convergence is $[-1, 1)$. By graphing several of the polynomials obtained by truncating the series, it is possible to visualize the convergence of the power series as a sequence of functions. Let's first plot the first through the fourth partial sums of the series—i.e., the first through the fourth degree polynomial approximations to the series—on the interval $[-2, 2]$.

```
polys = Table[Sum[(x^k / k), {k, 1, n}], {n, 1, 4}]
```

$$\left\{x,\ x+\frac{x^2}{2},\ x+\frac{x^2}{2}+\frac{x^3}{3},\ x+\frac{x^2}{2}+\frac{x^3}{3}+\frac{x^4}{4}\right\}$$

```
Plot[Evaluate[polys], {x, -2, 2}, PlotRange → {-2, 5}]
```

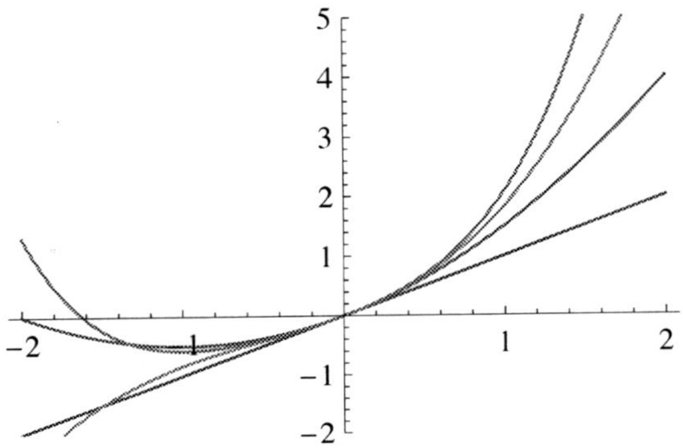

What we should notice here is that the graphs are very close to each other near $x = 0$. Now let's graph the fifth, tenth, fifteenth, and twentieth degree approximations to the series on the interval $[-1.5, 1.5]$.

```
Plot[Evaluate[Table[Sum[(x^k / k), {k, 1, n}], {n, 5, 20, 5}]],
  {x, -1.5, 1.5}, PlotRange → {-2, 5}]
```

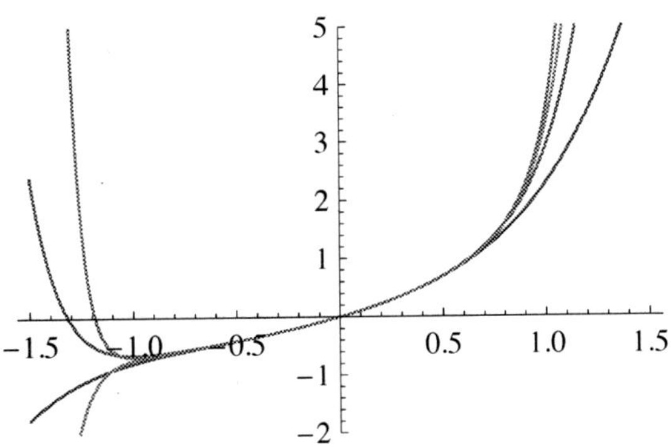

Here we see that the graphs are very close on a wider interval. Plotting the 100th and 101st degree approximations gives us a good picture of how the graph of the power series $f(x)$ would look.

```
Plot[Evaluate[{Sum[x^k/k, {k, 1, 100}], Sum[x^k/k, {k, 1, 101}]}], {x, -1.2, 1.1}, PlotRange → {-2, 5}]
```

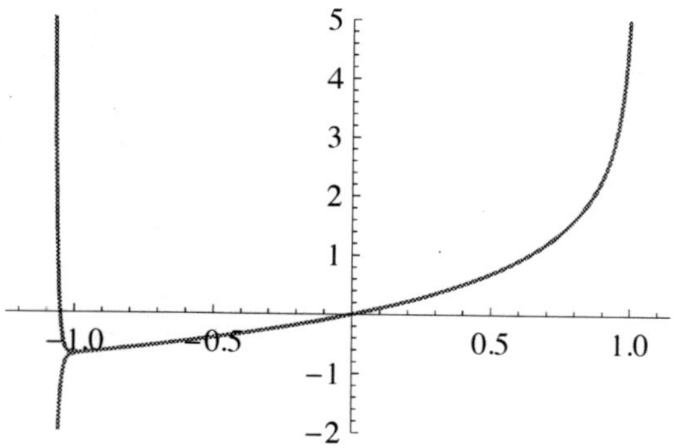

Note that the graph indicates that $f(x) \to \infty$ as $x = 1^-$ and that $f(x)$ has a finite limit as $x \to -1^+$. These facts are consistent with our previous observation that the interval of convergence is $[-1, 1)$.

■ Example 8.3.2

Again consider the power series

$$f(x) = \sum_{k=0}^{\infty} \frac{x^k}{k}$$

Let's try to determine the "**closed form**" of $f(x)$; that is, let's try to express $f(x)$ in a form that does not involve an infinite summation. The key here is to notice the possibility that $f(x)$ is related to the geometric series $\sum x^k$, which we know has the closed form $1/(1-x)$ for $-1 < x < 1$. In fact, by computing the derivative of f (term-wise), we see that

$$f'(x) = \sum_{k=1}^{\infty} x^{k-1} = \sum_{n=0}^{\infty} x^n = \frac{1}{1-x} \text{ for } -1 < x < 1.$$

Also, $f(0) = 0$. Therefore we can conclude that

$$f(x) = -\ln(1-x) \text{ for } -1 \le x < 1.$$

Let's look at the graph of this function, together with, say, the degree-twenty partial sum of the power series for comparison. The graph on the left is over the interval of convergence $-1 \le x \le 1$, and the graph on the right is over the wider interval $-1.3 \le x \le 1.3$.

```
P20[x_] := Sum[x^k / k, {k, 1, 20}]
```

```
GraphicsGrid[
 {{Plot[{P20[x], -Log[1 - x]}, {x, -1, 1}, PlotRange → {-1, 6}],
   Plot[{P20[x], -Log[1 - x]}, {x, -1.3, 1.3}, PlotRange → {-1, 6}]}}]
```

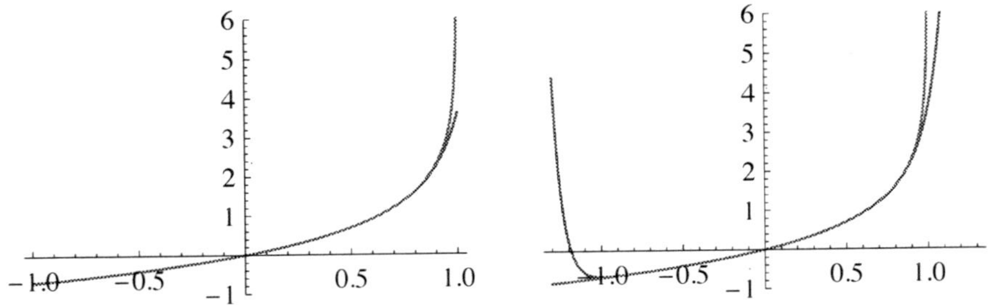

▪ Example 8.3.3

Consider the power series

$$f(x) = \sum_{k=1}^{\infty} (-1)^k x^{2k}.$$

The ratio test reveals that the series converges (absolutely) for $|x| < 1$ and diverges for $|x| > 1$. It also clearly diverges if $x = \pm 1$. So let's first plot the second through the fifth partial sums on the interval $[-1.5, 1.5]$.

```
(polys = Table[Sum[(-1)^k x^(2 k), {k, 0, n}], {n, 1, 5}]) // TableForm
```

$$\left(\text{polys} = \text{Table}\left[\sum_{k=0}^{n} (-1)^k x^{2k},\ \{n, 1, 5\}\right]\right) \text{// TableForm}$$

```
1 - x^2
1 - x^2 + x^4
1 - x^2 + x^4 - x^6
1 - x^2 + x^4 - x^6 + x^8
1 - x^2 + x^4 - x^6 + x^8 - x^10
```

```
Plot[Evaluate[polys], {x, -1.5, 1.5}, PlotRange → {-2, 3}]
```

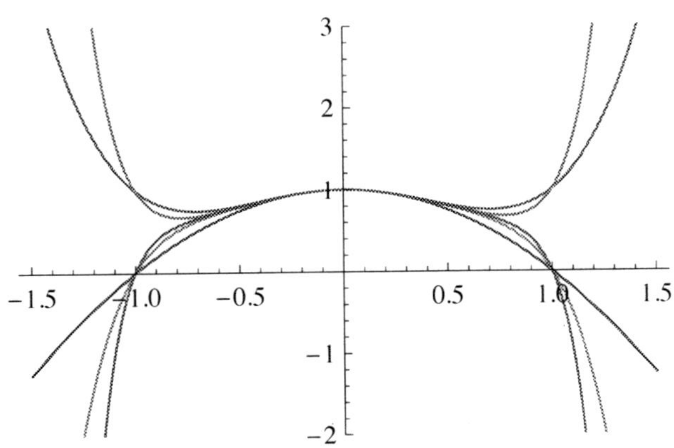

From what we know about the geometric series, it is not difficult to see the closed form:

$$\sum_{k=1}^{\infty}(-1)^k x^{2k} = \frac{1}{1+x^2} \quad \text{for } -1 \leq x \leq 1\,.$$

So let's plot the degree-twenty partial sum together with $y = \left(1+x^2\right)^{-1}$, again on the interval $[-1.5, 1.5]$.

$$\textbf{Plot}\left[\left\{\frac{1}{1+x^2}, \sum_{k=0}^{20}(-1)^k x^{2k}\right\}, \{x, -2, 2\}, \textbf{PlotRange} \to \{-.5, 2\}\right]$$

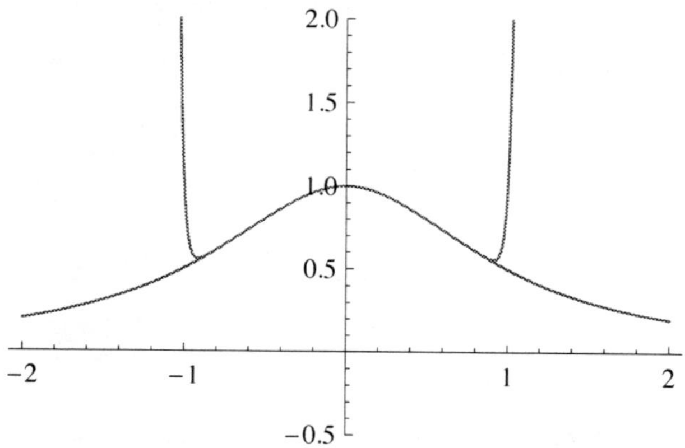

This graph does a very good job of illustrating the convergence of the partial sums of a power series. The function represented by the power series is defined and bounded for all x, but the power series converges to it only on the power series's interval of convergence.

◆ Exercises

For each of the following power series,

a) find the interval of convergence with the ratio test;

b) plot the partial sums of degree 0, 1, …, 4, and 10 on an appropriate interval;

c) find the closed form of $f(x)$ and plot it on f's interval of convergence.

1. $f(x) = \sum_{k=0}^{\infty}(-1)^k x^{3k}$
2. $f(x) = \sum_{k=0}^{\infty}\frac{(-1)^k}{k} x^k$
3. $f(x) = \sum_{k=1}^{\infty}\frac{(-1)^{k+1}}{k} x^{2k}$
4. $f(x) = \sum_{k=0}^{\infty}\frac{(-1)^k}{2k+1} x^{2k+1}$

8.4 Taylor Polynomials and Series

The nth degree **Taylor polynomial** about a number $x = a$ for a function f that is n-times differentiable in an open interval containing a is given by

$$T_n(x) = \sum_{k=0}^{n} \frac{f^{(k)}(a)}{k!}(x-a)^k,$$

where $f^{(k)}$ denotes the kth derivative of f. If f has derivatives of all orders in an open interval containing $x = a$, then its Taylor polynomials are partial sums of the power series

$$\sum_{k=0}^{\infty} \frac{f^{(k)}(a)}{k!}(x-a)^k,$$

which we call the **Taylor series** for f about $x = a$. When $a = 0$, this series is

$$\sum_{k=0}^{\infty} \frac{f^{(k)}(0)}{k!}x^k,$$

which is often called the **MacLaurin series** of f.

▪ Example 8.4.1

In this example we will construct Taylor polynomials "from scratch." Consider the function

```
f[x_] := (x^3 + 4 x^2 + 1)/(x^2 + 1);
Plot[f[x], {x, -1.5, 1.5}, PlotRange → {-1, 4}]
```

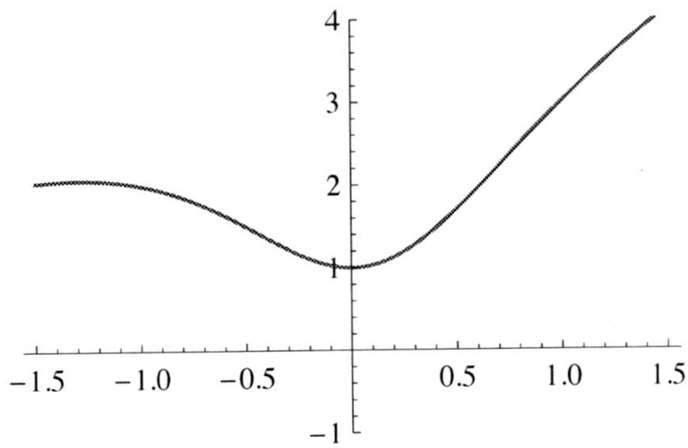

Let's find and plot the degree-two, -five, and -eight Taylor polynomials of f about $x = 0$. Toward that end, we may as well compute the first eight derivatives of f all at once:

```
(derivs = Table[D[f[x], {x, k}], {k, 0, 8}] // Simplify) // TableForm
```

$$\frac{1+4x^2+x^3}{1+x^2}$$
$$\frac{x\left(6+3x+x^3\right)}{\left(1+x^2\right)^2}$$
$$-\frac{2\left(-3-3x+9x^2+x^3\right)}{\left(1+x^2\right)^3}$$
$$\frac{6\left(1-12x-6x^2+12x^3+x^4\right)}{\left(1+x^2\right)^4}$$
$$-\frac{24\left(3+5x-30x^2-10x^3+15x^4+x^5\right)}{\left(1+x^2\right)^5}$$
$$\frac{120\left(-1+18x+15x^2-60x^3-15x^4+18x^5+x^6\right)}{\left(1+x^2\right)^6}$$
$$-\frac{720\left(-3-7x+63x^2+35x^3-105x^4-21x^5+21x^6+x^7\right)}{\left(1+x^2\right)^7}$$
$$\frac{5040\left(1-24x-28x^2+168x^3+70x^4-168x^5-28x^6+24x^7+x^8\right)}{\left(1+x^2\right)^8}$$
$$-\frac{40320\left(3+9x-108x^2-84x^3+378x^4+126x^5-252x^6-36x^7+27x^8+x^9\right)}{\left(1+x^2\right)^9}$$

Evaluating these derivatives at $x = 0$, we find

```
derivsAt0 = derivs /. x → 0
```

{1, 0, 6, 6, -72, -120, 2160, 5040, -120 960}

Now, the cubic Taylor polynomial is

$$\mathtt{T3[x\_]} = \sum_{k=0}^{3} \frac{\mathtt{derivsAt0[[k+1]]}}{k!} x^k$$

$1 + 3x^2 + x^3$

Plotting this along with f, we see

```
Plot[{f[x], T3[x]}, {x, -1.5, 1.5}, PlotRange → {-1, 4}]
```

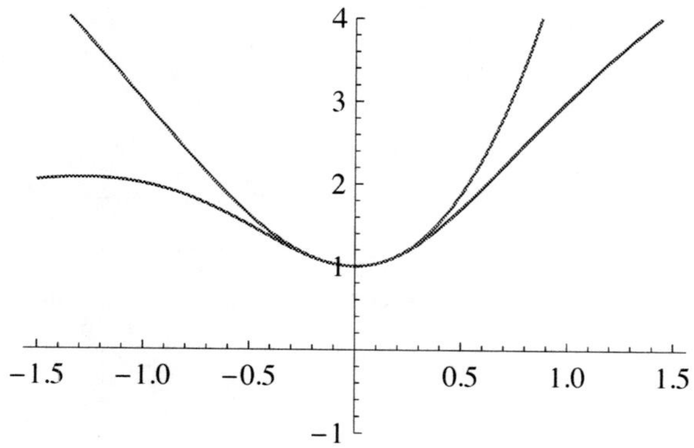

The fifth-degree Taylor polynomial is

```
T5[x_] = Sum[derivsAt0[[k + 1]]/k! x^k, {k, 0, 5}]
Plot[{f[x], T5[x]}, {x, -1.5, 1.5}, PlotRange → {-1, 4}]
```

$$T5[x\_] = \sum_{k=0}^{5} \frac{derivsAt0[[k+1]]}{k!} x^k$$

$1 + 3 x^2 + x^3 - 3 x^4 - x^5$

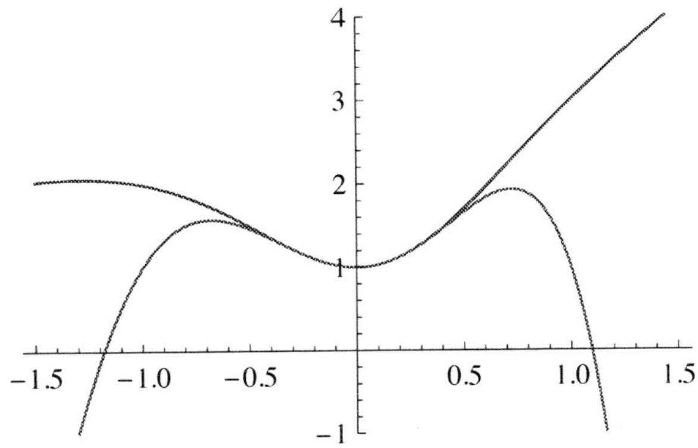

The eighth-degree Taylor polynomial is

$$T8[x\_] = \sum_{k=0}^{8} \frac{derivsAt0[[k+1]]}{k!} x^k$$

```
Plot[{f[x], T8[x]}, {x, -1.5, 1.5}, PlotRange → {-1, 4}]
```

$1 + 3 x^2 + x^3 - 3 x^4 - x^5 + 3 x^6 + x^7 - 3 x^8$

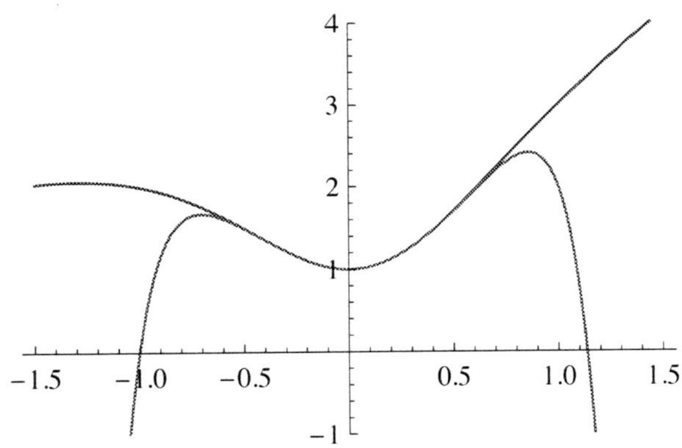

■ Example 8.4.2

In this example we will use *Mathematica*'s `Series` command to construct our series. Consider the function

$$f[x\_] := 1 \Big/ \sqrt{x^4 + x^3 + 1}$$

whose graph is as follows:

```
Plot[f[x], {x, -1.5, 1.5}, PlotRange → {0, 1.3}]
```

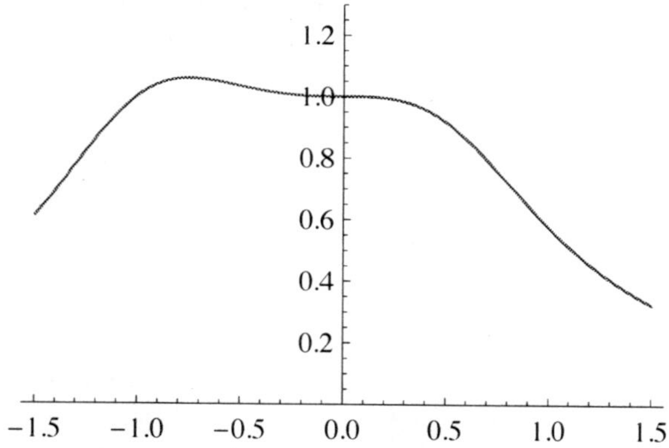

Let's first look at the cubic Taylor polynomial.

```
T3[x_] = Series[f[x], {x, 0, 3}]
```

$$1 - \frac{x^3}{2} + O[x]^4$$

Notice the $O[x]^4$ term. This indicates that the difference between the cubic Taylor polynomial and the function f is *on the order of* x^4; that is, the error behaves like a constant times x^4 as x approaches zero. We needn't dwell on this issue at this point; the important thing is that we must get rid of the $O[x]^4$ term before we can work with the cubic Taylor polynomial. This is done by applying `Normal` to the result of `Series`.

```
T3[x_] = Series[f[x], {x, 0, 3}] // Normal
```

$$1 - \frac{x^3}{2}$$

Now here's the plot:

```
Plot[{f[x], T3[x]}, {x, -1.5, 1.5}, PlotRange → {0, 1.3}]
```

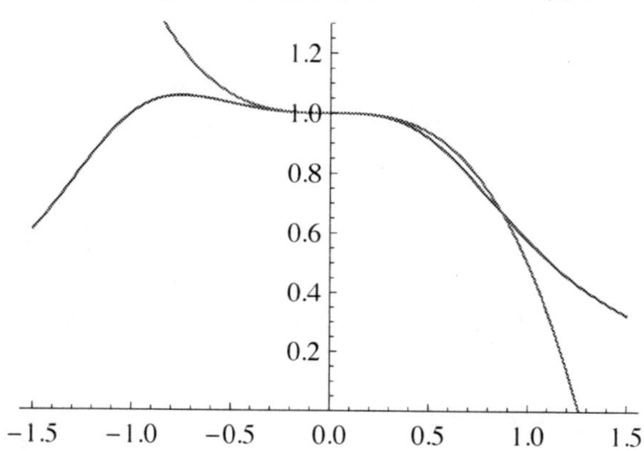

The following are the sixth degree Taylor polynomial and the resulting plot.

```
T6[x_] = Series[f[x], {x, 0, 6}] // Normal
Plot[{f[x], T6[x]}, {x, -1.5, 1.5}, PlotRange → {0, 1.3}]
```

$$1 - \frac{x^3}{2} - \frac{x^4}{2} + \frac{3\,x^6}{8}$$

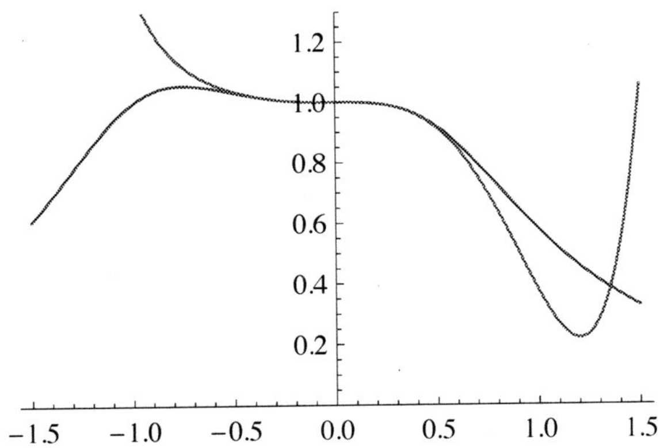

And now the ninth degree Taylor polynomial and the resulting plot:

```
T9[x_] = Series[f[x], {x, 0, 9}] // Normal
Plot[{f[x], T9[x]}, {x, -1.5, 1.5}, PlotRange → {0, 1.3}]
```

$$1 - \frac{x^3}{2} - \frac{x^4}{2} + \frac{3\,x^6}{8} + \frac{3\,x^7}{4} + \frac{3\,x^8}{8} - \frac{5\,x^9}{16}$$

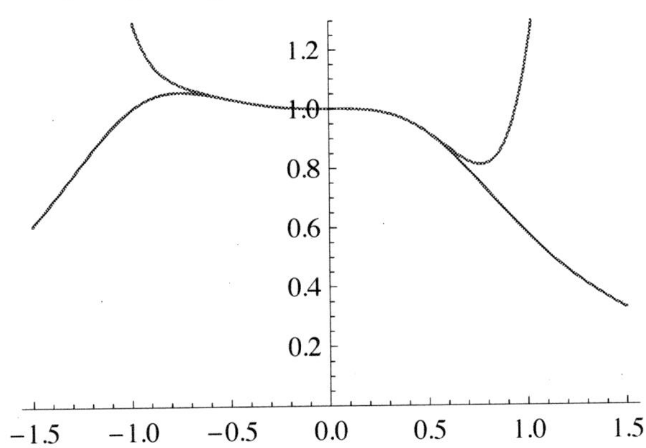

■ The Remainder Term

As the preceeding example show, Taylor polynomials provide an effective way of approximating a function locally about a given point $x = a$. The **remainder** $R_n(x)$ in the approximation of $f(x)$ by $T_n(x)$ is defined by

$$R_n(x) = f(x) - T_n(x).$$

According to **Taylor's Theorem**—provided that f is $(n+1)$-times differentiable on some interval containing a and x—the remainder satisfies

$$R_n(x) = \frac{f^{(n+1)}(c_x)}{(n+1)!}(x-a)^{n+1}$$

where c_x is some number between x and a.

■ Example 8.4.3

Approximate $\sin 0.3$ *by computing the value of the cubic Taylor polynomial about* $x = 0$ *at* 0.3. *Then use Taylor's Theorem to estimate the error.*

We begin by computing $T_3(x)$ and its value at $x = 0.3$.

```
T3[x_] = Series[Sin[x], {x, 0, 3}] // Normal
T3[.3]
```

$$x - \frac{x^3}{6}$$

```
0.2955
```

According to Taylor's theorem, the remainder is

$$\frac{\text{Sin}[c]}{4\,!}\,.3^4$$

```
0.0003375 Sin[c]
```

for some c between 0 and .3. The error is the absolute value of the remainder, which in this case is the same as the remainder, since $\sin c \geq 0$. Since we know nothing about c other than $0 \leq c \leq .3$, we need to produce an (over) estimate, i.e., an *upper bound*, on the error. Since $\sin x$ is increasing for $0 \leq x \leq .3$, it follows that

$$0.0003375 \sin c \leq 0.0003375 \sin 0.3.$$

We also know that $\sin 0.3 < 0.3$. (Why?) This gives us the error estimate

$$|R_3(.3)| < 0.0003375\,(.3) = 0.00010125.$$

A "sharper" error estimate can be found by realizing that the fourth degree Taylor polynomial for $\sin x$ about $x = 0$ is exactly the same as the cubic Taylor polynomial about $x = 0$.

```
T4[x_] = Series[Sin[x], {x, 0, 4}] // Normal
```

$$x - \frac{x^3}{6}$$

Therefore, we can estimate the error using $R_4(.3)$:

$$\frac{\text{Cos}[c]}{5\,!}\,.3^5$$

```
0.00002025 Cos[c]
```

The best upper bound for $\cos c, 0 \le c \le .3$, is 1. Therefore, we have the error estimate

$$|R_4(.3)| \le 0.00002025.$$

For the sake of comparison, let's compute the *actual* error:

```
Abs[Sin[.3] - T3[.3]]
```

```
0.0000202067
```

So our error estimate is quite good indeed.

Because of Taylor's Theorem, if we want to approximate $f(x)$ over an interval $[a-r, a+r]$, we have the *uniform* error estimate

$$|R_n(x)| \le \frac{M\,r^{n+1}}{(n+1)!} \quad \text{for } a-r \le x \le a+r,$$

where M is the maximum value of $\left|f^{(n+1)}(x)\right|$ on the interval $[a-r, a+r]$.

■ Example 8.4.4

Find a polynomial approximation to $f(x) = \cos x$ *uniformly on the interval* $[0, \pi/2]$ *with an error of not more than* 0.00005.

Since f and all of its derivatives have values between -1 and 1, we can take $M = 1$. So for any x in $[0, \pi/2]$ we have

$$|R_n(x)| \le \frac{(\pi/4)^{n+1}}{(n+1)!}.$$

Thus we want to find the least value of n such that

$$\frac{(\pi/4)^{n+1}}{(n+1)!} \le .00005.$$

A table of values indicates that $n = 8$ provides the desired error bound.

```
Table[{n, (π/4.)^(n+1)/(n+1)!}, {n, 4, 8}] // TableForm
```

4	0.00249039
5	0.000325992
6	0.0000365762
7	3.59086×10^{-6}
8	3.13362×10^{-7}

So the polynomial approximation we're looking for is

```
T6[x_] = Series[Cos[x], {x, π/4, 6}] // Normal // N // Expand
```

$$1.00002 - 0.000202346\,x - 0.499246\,x^2 - 0.00154123\,x^3 + 0.0435157\,x^4 - 0.00126455\,x^5 - 0.000982093\,x^6$$

The graph of $T_6(x)$ on $[0, \pi/2]$ is virtually identical to that of $\cos x$.

```
Plot[{T6[x], Cos[x]}, {x, 0, π/2}]
```

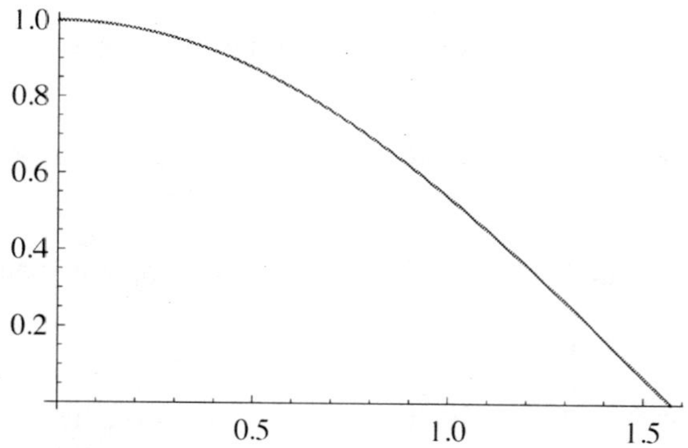

Plotting $T_6(x)$ together with $\cos x$ on a wider interval provides a better picture of the local nature of the polynomial approximation.

```
Plot[{T6[x], Cos[x]}, {x, -5, 6}, PlotRange → {-2, 2}]
```

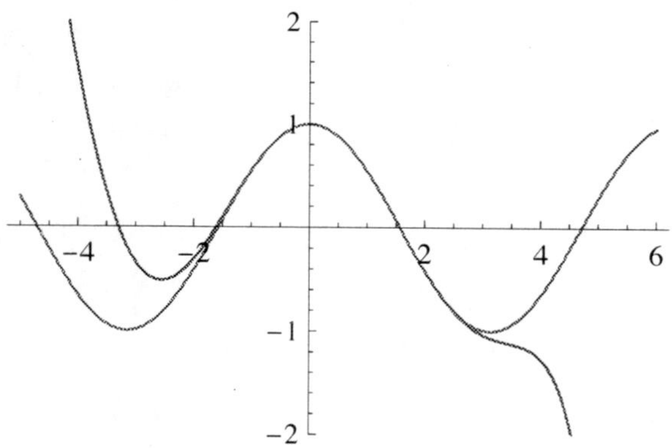

◆ Exercises

1. Find the fifth- and tenth-degree Taylor polynomials for $f(x) = \cos x$ about $x = \pi/2$. Then plot them both along with $\cos x$ on the interval $[-4, 7]$.
2. Find the fifth- and tenth-degree Taylor polynomials for $f(x) = \cos(1 - e^x)$ about $x = 0$ and plot them along with $\cos(1 - e^x)$ on the interval $[-2, 3]$.
3. Find the fifth- and tenth-degree Taylor polynomials for $f(x) = \left(1 + x^2\right)^{-1}$ about $x = 0$ and plot them along with $\left(1 + x^2\right)^{-1}$ on the interval $[-2, 2]$.
4. Find a polynomial approximation to $f(x) = e^x$ on the interval $[-2, 2]$ with an error of not more than 0.005. Then plot both the polynomial approximation and e^x on $[-4, 4]$.

5. Find a polynomial approximation to $f(x) = \ln x$ on the interval $[1/2, 3/2]$ with an error of not more than 0.005. Then plot both the polynomial approximation and $\ln x$ on the interval $[0.01, 4]$.

6. Find the third-, fourth-, and fifth-degree Taylor polynomials for $f(x) = x^3 + x^2 + x + 1$ about $x = 0$ and also about $x = 1$. What do you observe?

7. Use the cubic Taylor polynomial for $e^{\sin x}$ to approximate $e^{\sin 0.25}$. Use Taylor's Theorem to estimate the error in the approximation.

8. Use the fourth-degree Taylor polynomial for $\sqrt{1+x^4}$ about $x = 0$ to approximate the integral $\int_0^1 \sqrt{1+x^4}\, dx$. Estimate the error in the approximation.

9. Use the fourth degree Taylor polynomial for $\cos x$ about $x = 0$ to find an approximate formula for the positive solution of $\cos x = x^2 + k$, in terms of k, for $k < 1$. For what values of k does this give a reasonably accurate solution?

10. What does Taylor's Theorem say when $n = 0$? That special case is known as what theorem? What about $n = 1$?

11. Consider again the function $f(x) = (x^4 + x^3 + 1)^{-1/2}$ in Example 8.4.2. Plot f along with its 199th degree Taylor polynomial on the interval $[-1, 1]$ (with `PlotRange→{0,1.3}`). Use the plot to estimate the radius of convergence of f's Taylor *Series* about $x = 0$.

9 Projects

9.1 Two Limits

Background: Basic familiarity with doing simple computations and plotting graphs in *Mathematica*.

1. Suppose that a certain plant is one meter tall, and its height increases continuously at a rate of 100% per year. How tall will the plant be one year later?

   a) Suppose that instead of growing continuously the plant has monthly instantaneous growth spurts, in each of which its height increases by $\frac{100}{12}$ %. Compute the plant's height after twelve such monthly growth spurts.

   b) Suppose that instead of growing continuously the plant has daily instantaneous growth spurts, in each of which its height increases by $\frac{100}{365}$ %. Compute the plant's height after 365 such daily growth spurts.

   c) Suppose that instead of growing continuously the plant has n instantaneous growth spurts during the year, in each of which its height increases by $\frac{100}{n}$ %. Express the year-end height of the plant as a function $h(n)$.

   d) Plot the graph of $h(n)$ for $1 \leq n \leq 500$. Describe the behavior of the graph and estimate the number that $h(n)$ approaches as n gets larger and larger. Now give an (approximate) answer to the original question.

2. Suppose that your friend asks for a loan of \$2500 and offers to pay you back according to the following schedule. He will pay you \$100 tomorrow, and on each day thereafter he will pay you 95% of the previous day's payment for the rest of your life—or until the daily payment amount is less than one penny. Should you agree to the deal?

   a) The amount of the loan that will have been paid back after n days is

   $$A(n) = 100 + 100\,(.95) + 100\,(.95)^2 + 100\,(.95)^3 + \cdots + 100\,(.95)^{n-1}.$$

   So define

   $$\texttt{a[n\_] := 100}\sum_{i=1}^{n-1}\texttt{.95}^{i}$$

   and then plot the graph of $A(n)$ for $0 \leq n \leq 30$. How much of the loan has been paid back after 30 days?

   b) The method used in (a) to compute $A(n)$ is very inefficient. Show that

   $$A(n) - .95\,A(n) = 100 - 100\,(.95)^n = 100\left(1 - .95^n\right)$$

and consequently that

$$A(n) = \frac{100\left(1 - .95^n\right)}{.05} = 2000\left(1 - .95^n\right).$$

Redefine `a[n]` using this simpler formula. Then plot its graph for $1 \le n \le 365$. What can you conclude about when the loan will be paid off?

c) After how many days will the daily payment become less than a penny? How much money you would get back if you did agree to the deal?

9.2 Computing π as an Area

Background: Basic trigonometry.

In this project, we will explore a method for computing π that was known to the ancient Greeks. It is sometime called the "method of exhaustion." (But with *Mathematica*'s help, hopefully *we* won't become exhausted. ☺) The method embodies the essence of calculus: *successively improved approximations*.

Let us suppose that we *define* π to be area inside a circle of radius 1. Then by approximating the area inside the circle, we approximate π. The way that we'll approximate this area is to compute the area of a regular polygon whose vertices lie on the circle.

1. Before we proceed with the computations, let's first create some graphics to illustrate what we're doing.

   a) Enter the function definition

   ```
   points[n_] := Table[{Cos[t], Sin[t]}, {t, 2 π / n, 2 π, 2 π / n}]
   ```

   This function will create a list of any number of equally space points around the unit circle. Test the function by entering

   b) Now enter

   ```
   ngon[n_] := Graphics[{Pink, Polygon[points[n]]}]
   ```

   which defines a function that produces a regular n-gon as a graphics object. Test the function by entering

   ```
   Show[ngon[6]]
   ```

   c) Our method for computing the area of a regular n-gon will involve dividing the region up into n triangles. Enter

   ```
   spokes[n_] :=
    Graphics[{Opacity[.5], Map[Line[{{0, 0}, #}] &, points[n]]}]
   ```

   to define a function that will produce a graphics object consisting of the edges of the triangles that don't lie on the n-gon. Test by entering

   ```
   Show[ngon[6], spokes[6]]
   ```

d) To add the circle to the picture, enter

```
circ = Graphics[{Blue, Disk[{0, 0}, 1]}]
```

and then

```
pic[n_] := Show[circ, ngon[n], spokes[n]]
```

We now have a function that will draw the full picture for any n. Test it by entering

```
pic[8]
```

e) Create an animation by entering

```
Animate[pic[n], {n, 3, 40, 1}]
```

The animation will show you that the n-gons provide better and better approximations to the circle as n increases.

2. a) Now we need a formula for the area of each n-gon. Using basic trigonometry, and noting that each triangle can be divided up into two congruent *right* triangles, show that each of the n triangles within the n-gon has area

$$\sin\frac{180^\circ}{n}\cos\frac{180^\circ}{n},$$

and consequently the area within the n-gon is

$$n\sin\frac{180^\circ}{n}\cos\frac{180^\circ}{n}.$$

(*Note*: We are using degree measure for angles because radian measure requires us to know π.)

b) Define an area function by entering

```
area[n_] := n Sin[180 Degree / n] Cos[180 Degree / n]
```

Then compute a table of values by entering

```
Join[Table[{n, N[area[n]]}, {n, 10, 100, 10}],
    Table[{n, N[area[n]]}, {n, 200, 2000, 100}]] // TableForm
```

3. Finally, let's add a feature to the `pic` function in #1. To include a label showing the area of the n-gon, enter the definition

```
areaLabel[n_] := Graphics[
  Text[StyleForm[N[area[n]], FontSize → 14], {.85, .9}]]
```

and then insert `areaLabel[n]` immediately after `ngon[n]` in the definition of `pic`. Redo the animation in #1e; then create a similar animation by entering

```
Animate[pic[2^n], {n, 3, 10, 1}, AnimationRepetitions → 1]
```

9.3 Lines of Sight

Background: Tangent lines; slope.

Ant Man is standing 100 feet away from a tall building and 12 feet away from a greenhouse with semicircular cross-section of radius 12 feet, as shown in the figure. Ant Man's eyes are 5.6 feet above the ground. How high above the ground is the lowest point he can see on the side of the building, and how high above the ground is the highest point he can see on the roof of the greenhouse?

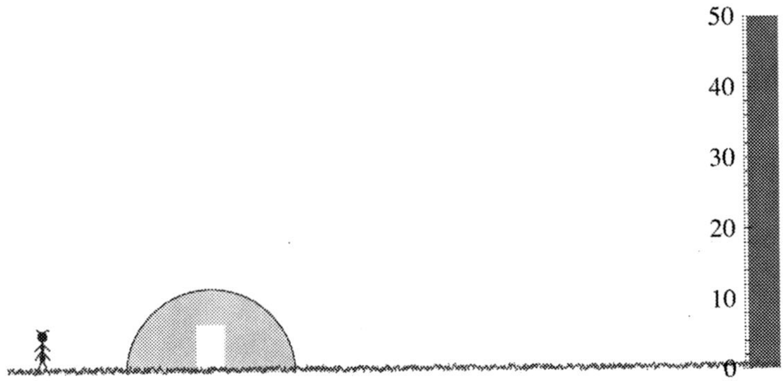

Follow the steps below to solve the problem—first graphically and then with calculus.

1. First define the function

```
f[x_] := √(144 - (x - 24)²)
```

to represent the roof of the greenhouse. Then create the view of the scene shown above by entering

```
greenhouse = Plot[f[x], {x, 12, 36},
  Filling → Axis, PlotRange → {{-5, 105}, {0, 50}},
  AspectRatio → Automatic, AxesOrigin → {99.5, 0}]

grass = Plot[RandomReal[], {x, -5, 105},
  PlotStyle → Darker[Green], PlotPoints → 20,
  PlotRange → {{-5, 105}, {0, 50}}, AspectRatio → Automatic,
  AxesOrigin → {99.5, 0}]
```

followed by

```
Show[antman = Graphics[{RGBColor[.5, .1, 0],
    Line[{{-1, 0}, {0, 2.2}, {0, 6}, {0, 2.2}, {1, 0}}],
    Line[{{-1, 3.5}, {0, 4.5}, {1, 3.5}}],
    Line[{{-1, 2.2}, {0, 3.5}, {1, 2.2}}],
    Line[{{-.75, 6.5}, {0, 6}, {.75, 6.5}, {.9, 6.4}}],
    Disk[{0, 5.5}, .75], Disk[{0, 3.9}, {.25, .8}],
    Disk[{0, 2.4}, {.4, .6}]}], ImageSize → 24]
```

```
bldg = Graphics[{Gray,
    Polygon[{{100.5, 0}, {100.5, 50}, {105, 50}, {105, 0}}]}];
door = Graphics[{White, Polygon[
     {{22, 0}, {22, 7}, {26, 7}, {26, 0}}]}];
```

and then

```
scene = Show[bldg, antman, greenhouse, door, grass,
  PlotRange → {{-5, 105}, {0, 50}},
  Axes → {False, True}, AxesOrigin → {100, 0}]
```

2. Ant Man's line of sight will be a line segment that begins at (0, 5.6) and ends either at a point on the greenhouse roof or on the side of the building. Let's use the segment's slope m as the parameter that determines where the end of the line of sight. Then, if the line of sight ends on the greenhouse roof, the equation

$$mx + 5.6 = \sqrt{144 - (x - 24)^2}$$

will have a real solution. If the line of sight ends on the side of the building, then that equation will have only complex solutions. For example, with $m = 0.1$, the equation can be solved as follows.

```
m = .1;
soln = x /. Solve[m x + 5.6 == f[x], x]
```

```
{14.2729, 32.143}
```

{14.2729, 32.143}

The first of the two solutions is the x-coordinate of the point where the line of sight hits the greenhouse roof. By plotting the line $mx + 5.6$ between 0 and that number, we get a picture of the line of sight.

```
Show[scene,
 Plot[m x + 5.6, {x, 0, soln[[1]]}, PlotStyle → {Dashed, Red}]]
```

With $m = 0.4$, the equation has only complex solutions:

```
m = .4;
soln = x /. Solve[m x + 5.6 == f[x], x]
```

```
{18.7586 - 6.89655 i, 18.7586 + 6.89655 i}
```

In this situation, we can plot the line $mx + 5.6$ over $0 \le x \le 100$ to see the line of sight.

```
Show[scene,
 Plot[m x + 5.6, {x, 0, 100}, PlotStyle → {Dashed, Red}]]
```

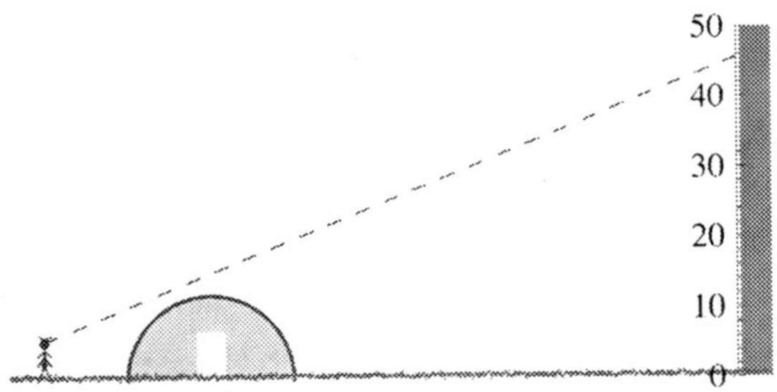

By repeating this process for different values of m, experimentally estimate the value of m for which the line of sight just grazes the greenhouse roof. Then give answers to the questions posed.

3. To solve the problem with calculus, first enter

```
tangent[x_, a_] = f[a] + f′[a] (x - a)
```

which gives the function whose graph is the tangent line to the greenhouse roof at $(a, f(a))$. Then, for the sake of illustration, plot that line over $0 \le x \le 100$ for $a = 18, 20, 22, 24$, showing the line along with the `scene` graphic from #2.

Find the x-coordinate of the point of interest on the roof of the greenhouse by solving the equation

```
eqn := tangent[0, a] == 5.6
```

Explain why this is the appropriate equation to solve. Then use the result to give answers to the questions posed.

9.4 Color-coded Graphs

Background: First and second derivatives of a function f and their relationship to geometric properties of the graph of f.

Our first goal is to create a plot in which the graph of a function is color-coded according to whether the function is increasing or decreasing. After that, we will color-code according to concavity and then according to both incresing/decreasing and concavity.

Given a function f, a related function that coincides with f wherever f is increasing and is undefined elsewhere is

```
fIncr[x_] := If[f′[x] > 0, f[x]]
```

Similarly, a related function that coincides with $f(x)$ wherever $f(x)$ is decreasing and is undefined elsewhere is

```
fDecr[x_] := If[f′[x] < 0, f[x]]
```

1. Enter the preceding definitions of `fIncr` and `fDecr` as well as

```
f[x_] := Sin[x] Cos[3 x];
a := 0; b := π;
```

Then plot separately the increasing and decreasing parts of the graph of f over $0 \le x \le \pi$.

```
Plot[fIncr[x], {x, a, b}, PlotRange → {{a, b}, Automatic}]
Plot[fDecr[x], {x, a, b}, PlotRange → {{a, b}, Automatic}]
```

2. Combine the two graphs with different colors—blue for increasing, orange for decreasing—by entering

```
Plot[{fIncr[x], fDecr[x]},
 {x, a, b}, PlotRange → {{a, b}, Automatic},
 PlotStyle → {{Darker[Blue], Thick}, {Orange, Thick}}]
```

3. Let's now add to the plot the graph of the derivative, color-coded to match the coloring of the graph of the function. Enter these definitions for the positive and negative "pieces" of the derivative:

```
derPos[x_] := If[f′[x] > 0, f′[x]]
derNeg[x_] := If[f′[x] < 0, f′[x]]
```

Now create the plot by entering

```
Plot[{fIncr[x], derPos[x], fDecr[x], derNeg[x]},
 {x, a, b}, PlotRange → {{a, b}, Automatic},
 PlotStyle → {{Darker[Blue], Thick},
   {Darker[Green], Thick}, {Orange, Thick}, {Red, Thick}}]
```

4. Modify the preceding procedure to color-code according to concavity.

5. Finally, let's color-code according to both concavity *and* whether the function is increasing or decreasing. There are four combinations, which are described by the following definitions.

```
fIncrUp[x_] := If[f′[x] > 0 && f″[x] > 0, f[x]]
fDecrUp[x_] := If[f′[x] < 0 && f″[x] > 0, f[x]]
fIncrDn[x_] := If[f′[x] > 0 && f″[x] < 0, f[x]]
fDecrDn[x_] := If[f′[x] < 0 && f″[x] < 0, f[x]]
```

Now enter the following, which assigns different colors to each of the four combinations.

```
Plot[{fIncrUp[x], fDecrUp[x], fIncrDn[x], fDecrDn[x]}, {x, a, b},
 PlotStyle → Table[{Hue[k, 1, .8], Thick}, {k, .25, 1, .25}]]
```

Finally, describe the meaning of each color in this coding.

6. Create color-coded plots of each of the three kinds from #3, #4, and #5 for each of the following functions on the specified interval.

a) $f(x) = (x^2 - 1)\,x^3, \ -1.1 \le x \le 1.1$

b) $f(x) = x^3\, e^{-x^2}, \ -3 \le x \le 3$

c) $f(x) = (1 - x)\,(x^2)^{1/3}, \ -.5 \le x \le 1.5$

d) $f(x) = \dfrac{x^3 - 10}{4 - x^2}, \ -7 \le x \le 7$

9.5 Designing an Oil Drum

Background: Optimization.

You work for a company that manufactures cylindrical steel drums that can be used to transport various petroleum products. Your assignment is to determine the dimensions (radius and height) of a drum that is to have a volume of 1 cubic meter while minimizing the cost the drum.

1. The cost of the steel used in making the drum is \$3 per square meter. The top and bottom of the drum are cut from squares, and all unused material from these squares is considered waste. The remainder of the drum is formed from a rectangular sheet of steel, assuming no waste. Ignoring all costs other than material cost, find the dimensions of the drum that will minimize the cost. *Suggestions*:

   a) Using the fact that the drum's volume is to be 1 cubic meter, express the height h of the drum in terms of its radius r. Then express the material cost first in terms of both r and h and then as a function of r alone. (Don't forget to include the waste resulting from cutting the top and bottom from squares.) Enter the cost as a function of r named `matCost`.

   b) Plot the graph of the material cost for $0 \le r \le 2$ (with `PlotRange→{0,100}`). Compute the derivative and use it to locate the value of r that minimizes the material cost. What is the corresponding height?

2. The drum has seams around the perimeter of its top and bottom, as well as a vertical seam where edges of the rectangular sheet are joined to form the lateral surface. In addition to the material cost, the cost of welding the seams is \$2 per meter. Find the dimensions of the drum that will minimize the cost of production. *Suggestions*:

   a) Add the seam-welding cost, in terms of r alone, to the material cost. Enter the result as a function named `prodCost`.

   b) Plot the graph of the production cost for $0 \le r \le 2$ (with `PlotRange→{0,100}`). Compute the derivative and use it to locate the value of r that minimizes the production cost. What is the corresponding height?

3. The cost of shipping each drum from your plant in Birmingham to an oil company in New Orleans depends upon both the surface area of the drum (which determines weight) and the sum of the diameter and height (which affects how many drums can be transported on one vehicle). This cost is estimated to be \$1 per square meter of surface area plus \$0.50 per meter of diameter plus height. Find the dimensions of the drum that will minimize the total cost of production and transportation. *Suggestions*:

   a) Add the transportation cost, in terms of r alone, to the production cost. Enter the result as a function named `totalCost`.

   b) Plot the graph of the total cost for $0 \le r \le 2$ (with `PlotRange→{0,100}`). Compute the derivative and use it to locate the value of r that minimizes the total cost. What is the corresponding height?

9.6 Optimal Location of a Water Treatment Plant

Background: Distance formula; curve fitting; the derivative; optimization.

Three towns—Appalachee, Hull, and Eastville—agree to share the cost of constructing a new water treatment plant that will supply all three towns with water. The plant will be located on a nearby river. The main concern in deciding where to locate the plant is the total cost of installing the supply pipelines to all three towns and building an access road from the plant to a nearby interstate highway. The map below shows the center of each town, the river, and the interstate highway. The grid lines on the map are 1 mile apart.

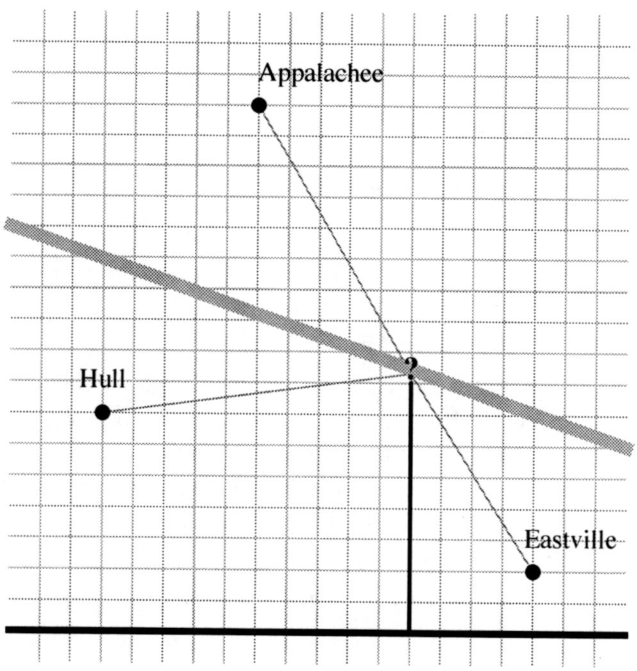

The cost of the pipeline to Eastville will be $15,000 per mile. The cost of installing pipelines to each of Appalachee and Hull will be $40,000 per mile, since they must go through heavily wooded areas. The cost of building the road will be $180,000 per mile. Assume that the supply pipelines extend to the centers of the three towns and ignore the width of the river.

1. Find the point on the river where the new water treatment plant should be located in order to minimize the total cost of installing the supply pipelines and building the road.

9.7 Newton's Method and a 1D Fractal

Background: Newton's Method.

Consider the simple cubic polynomial

$$f(x) = x^3 - x\,,$$

which has three distinct real zeros: 0 and ± 1. Our purpose here is to explore the question of which initial guesses x_0 cause Newton's Method to converge to each of these three zeros.

1. Plot the graph of f by entering

```
f[x_] := x (x² - 1);
Plot[{f[x], f[.472] + f'[.472] (x - .472)}, {x, -1.5, 1.5}]
```

For future reference, find the values of x where the local extrema occur.

2. Define the Newton iteration function by entering

```
newtStep[x_] = x - f[x]/f'[x] // Simplify // N
```

Then do the following sample calculation.

```
NestList[newtStep, 2, 6]

{2, 1.45455, 1.15105, 1.02533, 1.00091, 1., 1.}
```

3. We are going to use *Mathematica*'s `FixedPoint` function to define a function named `newton`, which will associate an initial guess x with the consequent root of f found by Newton's Method. The `SameTest` option lets us to specify a stopping criterion for the iteration. (The default stopping criterion is *identical* consecutive iterates.) For our purposes here, it will suffice to stop the iteration when consecutive iterates are merely *close* to each other. An appropriate function for testing closeness is given by

```
closeQ[xOld_, xNew_] := Abs[xOld - xNew] < 5 * 10^-4
```

Note, for example:

```
closeQ[1.1, 1.099]
closeQ[1.1, 1.0996]

False

True
```

Define the function `newton` as follows. The third argument limits the number of iterations to 30.

```
newton[x_] := FixedPoint[newtStep, x, 30, SameTest → closeQ]
```

Then do the following sample calculations.

```
{newton[.3], newton[.4], newton[.5], newton[.6], newton[.7]}

{2.77334 × 10^-27, 1.84213 × 10^-12, -1., 1., 1.}
```

4. The following is the graph of `newton` for $-1.2 \le x \le 1.2$. Note that the values of this function correspond to the roots of f.

```
Plot[newton[x], {x, -1.2, 1.2}]
```

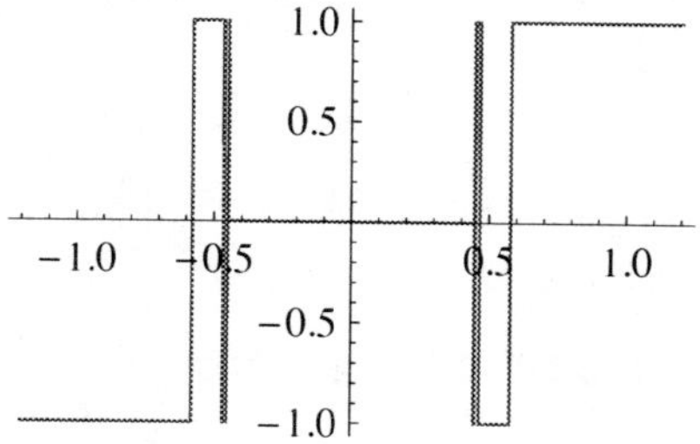

   a) There is an interval of the form (a, ∞) that contains 1 and has the property that the Newton iterates converge to 1 whenever the initial guess x_0 is in that interval. Considering the graph of f, the value of a is geometrically obvious. What is it?

   b) Let a have its value from (a). There is an interval of the form $(a-h, a)$ with the property that the Newton iterates converge to -1 whenever the initial guess x_0 is in that interval. If $a-h$ were used as x_0, what would be the resulting value of x_1? Use this to write and solve an equation for h. What then are the endpoints of the interval $(a-h, a)$? Notice that there is an analogous interval $(-a, -a+h)$ containing initial guesses for which the Newton iterates converge to 1.

   c) There is an interval of the form $(-r, r)$ with the property that the Newton iterates converge to 0 whenever the initial guess x_0 is in that interval. If $-r$ were used as the initial guess x_0, what would be the resulting value of x_1? Use this to write and solve an equation for r. What then are the endpoints of the interval?

   Because of symmetry, we may as well restrict our further investigations to $x \ge 0$.

```
Plot[newton[x], {x, 0, 1.}]
```

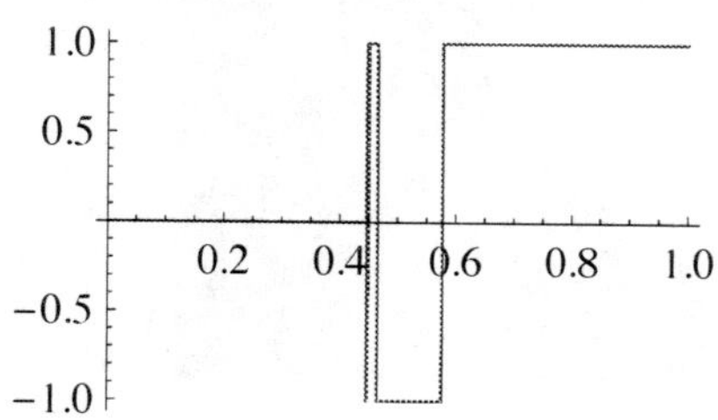

5. Enter the following, thereby gradually zooming in on the most interesting part of the graph.

```
GraphicsGrid[
 {{Plot[newton[x], {x, .44, .46}, PlotStyle → Thickness[.004]]},
  {Plot[newton[x], {x, .447, .4475}, PlotStyle → Thickness[.004]]},
  {Plot[newton[x],
    {x, .44721, .44722}, PlotStyle → Thickness[.004]] }}]
```

   Describe in some detail what you observe. How would you describe the set of all x_0 for which the Newton iterates converge to 1?

■ Extra Cool Stuff

An interesting color-coded graph roughly corresponding to those in #5 is produced by the following. In the resulting plot, blue indicates convergence to 0, green indicates convergence to −1, and red indicates convergence to 1.

```
Show[Graphics[
  Raster[{.67 + .33 Table[newton[x], {x, .44, .63, 10^-3}]},
   ColorFunction → Hue]], AspectRatio → .02]
```

A far more interesting picture is obtained by expanding our initial guesses to the complex plane. In order to speed up the calculation, we'll create a *compiled* version of `newton`:

```
newtC = Compile[{{x, _Complex}},
  Abs[1 + FixedPoint[(2 #^3/(3 #^2 - 1)) &, x, 30,
    SameTest → ((Abs[#1 - #2] < 5 * 10^-4) &)]]]
```

Now we use `DensityPlot` to create a well-known *fractal image*.

```
DensityPlot[newtC[x + y i], {x, .44, .63}, {y, -.05, .05},
 AspectRatio → Automatic, PlotPoints → 100,
 ColorFunction → (Lighter[Hue[2 # / 3]] &),
 FrameLabel → {Re[z], Im[z]}, RotateLabel → False]
```

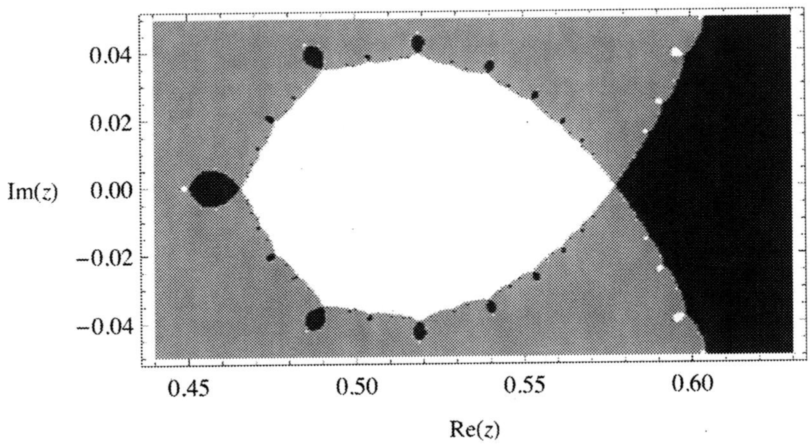

9.8 The Vertical Path of a Rocket

Background: Antidifferentiation; velocity and acceleration; rectilinear motion.

A small rocket has a mass of 100 kg, including 30 kg of fuel. Its engine is ignited at time $t = 0$ and produces a constant thrust of 980 Newtons as it burns fuel for 30 seconds at a constant rate of 1 kg/sec, propelling the rocket upward on a vertical path.

1. a) Express the mass m of the rocket, including fuel, as a function of t, for $t \geq 0$.

   b) Define a function `m1[t_]` that gives the mass of the rocket for $0 \leq t \leq 30$.

2. Let $y(t)$ denote the height (in meters) and $v(t)$ the velocity of the rocket at time t seconds. If we ignore air resistance, Newton's Second Law gives the equation

$$m(t)\, v'(t) = -\tfrac{98}{10}\, m(t) + 980\,.$$

   a) Find the velocity of the rocket for $0 \leq t \leq 30$ by computing

$$\texttt{v1[t\_]} = \int_0^t \left(\frac{980}{\texttt{m1[s]}} - \frac{98}{10} \right) ds$$

   b) Find the height of the rocket for $0 \leq t \leq 30$ by computing

$$\texttt{y1[t\_]} = \int_0^t \texttt{v1[s]}\, ds$$

   c) Compute the height and the velocity of the rocket at the instant when it runs out of fuel.

3. After the rocket runs out of fuel, the problem becomes a simple one concerning the height of a projectile for which the only acceleration comes from gravity, and the initial height and velocity are known. Letting T be the time when the rocket hits the ground, we have

```
y2[t_] := -4.9 (t - 30)^2 + v1[30] (t - 30) + y1[30]
```

   for $30 \leq t \leq T$ describing the height during the rocket's "free-fall" after it runs out of fuel. Enter the above definition of `y2`, and then plot the graph of the complete height function

```
height[t_] :=
 Which[0 ≤ t ≤ 30, y1[t], y2[t] > 0, y2[t], y2[t] ≤ 0, 0]
```

4. Find the maximum height attained by the rocket and the time T at which the rocket hits the ground.

9.9 Otto the Daredevil

Background: Antidifferentiation; velocity and acceleration; rectilinear motion.

Situation: Otto is an unusual daredevil. He jumps off of tall buildings with a small computer-controlled jetpack strapped to his back. The jetpack carries a small amount fuel—just enough to last for 10 seconds. The acceleration it provides is 14.4 m/sec^2. Moreover, the jetpack can be switched on and then off only once. Thus, on each jump, Otto's goal is to program the jetpack to switch on and then off at precisely the right moments so that he lands with zero velocity. (Throughout this project, ignore air resistance and assume that $g = 9.8$ m/sec^2.)

Preliminaries: Let's denote the height of the building by h. Then, between the time he jumps ($t = 0$) and the time the jetpack switches on, Otto's height and velocity are

```
y1[t_] := -9.8 t² / 2 + h
v1[t_] = ∂t y1[t];
```

Let s_1 and s_2 denote the times that Otto programs his jet pack to switch on and off, respectively. Then, for $s_1 \le t \le s_2$, Otto's height and velocity are

```
y2[s1_, t_] := (14.4 - 9.8) (t - s1)² / 2 + v1[s1] (t - s1) + y1[s1];
v2[s1_, t_] = ∂t y2[s1, t];
```

If he does not land by time s_2, then for t between s_2 and the time he does land, his height and velocity are

```
y3[{s1_, s2_}, t_] := -4.9 (t - s2)² + v2[s1, s2] (t - s2) + y2[s1, s2];
v3[{s1_, s2_}, t_] = ∂t y3[{s1, s2}, t];
```

All of the above can be pieced together as follows to get a complete height function for $t \ge 0$:

```
height[{s1_, s2_}, t_] := Which[t < s1, y1[t], t < s2, y2[s1, t],
  y3[{s1, s2}, t] > 0, y3[{s1, s2}, t], y3[{s1, s2}, t] ≤ 0, 0]
```

1. By graphical experimentation, determine how Otto should program his jetpack (i.e., find s_1, s_2) for:

   a) a 100 meter jump; b) a 200 meter jump.

   Suggestion: Begin by entering

   ```
   h = 100; Plot[height[{2, 7}, t], {t, 0, 11}, PlotRange → {0, h}]
   Clear[h]
   ```

2. How tall is the tallest building from which Otto should jump if he insists upon landing with zero velocity? (Remember that the jetpack has only enough fuel to last 10 seconds.) First obtain an approximate solution by graphical experimentation. Then find the solution by solving the appropriate pair of equations for s_1 and h. (*Hint*: You want the height and the velocity each to be zero at the time when the jetpack runs out of fuel.)

3. How tall is the tallest building from which Otto should jump if he doesn't mind landing with a velocity of -10 m/sec?

9.10 Helping *Mathematica* Integrate

Background: Antidifferentiation; the substitution rule.

One of the most extraordinary capabilities of *Mathematica* is its ability to compute antiderivatives of elementary functions. Indeed, if *Mathematica* is not able to antidifferentiate a particular function, then, *most likely*, that function does not have an antiderivative expressible in terms of other elementary functions or any of the many *special functions* that typically arise. Such functions are easy to come by; for example,

$$\int x^{\sqrt{x}}\, dx$$

$$\int x^{\sqrt{x}}\, dx$$

However, there are many functions with elementary antiderivatives that *Mathematica* cannot find. In spite of the fact that *Mathematica* is well-aware of the substitution rule,

$$\int f'[u[x]]\, u'[x]\, dx$$

```
f[u[x]]
```

such a failure usually happens because *Mathematica* does not recognize the particular substitution that makes antidifferentiation possible. This is a task at which humans are still far more capable than machines.

For example, consider the evaluation of

$$\int \left(1 + \frac{1}{2\sqrt{x}}\right)\left(x + \sqrt{x}\right)\sqrt{1 + \left(x + \sqrt{x}\right)^4}\, dx.$$

We quickly find that a direct approach with *Mathematica* is not successful:

$$\int \left(1 + \frac{1}{2\sqrt{x}}\right)\left(x + \sqrt{x}\right)\sqrt{1 + \left(x + \sqrt{x}\right)^4}\, dx$$

$$\int \left(1 + \frac{1}{2\sqrt{x}}\right)\left(\sqrt{x} + x\right)\sqrt{1 + \left(\sqrt{x} + x\right)^4}\, dx$$

However, the integrand suggests that we at least *try* the substition $u = x + \sqrt{x}$. In fact, that substitution works, because if

```
u[x_] := x + √x
```

then

```
u'[x] // Simplify
```

$$1 + \frac{1}{2\sqrt{x}}$$

So with

```
f[u_] := u Sqrt[1 + u^4]
```

the integrand is equal to

```
f[u[x]] u'[x] // Simplify
```

$$\frac{1}{2}\left(1+\sqrt{x}\right)\left(1+2\sqrt{x}\right)\sqrt{1+\left(\sqrt{x}+x\right)^4}$$

Consequently, we need only antdifferentiate f,

```
intf = ∫f[u] du // Simplify
```

$$\frac{1}{4}\left(u^2\sqrt{1+u^4}+\text{ArcSinh}\left[u^2\right]\right)$$

and then compose the result with u:

```
antiD[x_] = intf /. u → u[x] // Simplify
```

$$\frac{1}{4}\left(\left(\sqrt{x}+x\right)^2\sqrt{1+\left(\sqrt{x}+x\right)^4}+\text{ArcSinh}\left[\left(\sqrt{x}+x\right)^2\right]\right)$$

We can now check the result:

```
antiD'[x] // Simplify // Factor
Simplify[% == (1 + 1/(2 Sqrt[x])) (x + Sqrt[x]) Sqrt[1 + (x + Sqrt[x])^4]]
```

$$\frac{1}{2}\left(1+\sqrt{x}\right)\left(1+2\sqrt{x}\right)\sqrt{1+\left(\sqrt{x}+x\right)^4}$$

```
True
```

1. Use an approach similar to that described above to assist *Mathematica* in evaluating each of the following.

a) $\displaystyle\int (\cos x - x\sin x)\,x\cos x\sqrt{1+x^3\cos^3 x}\;dx$

b) $\displaystyle\int (1+\ln x)\,x\ln x\sqrt{1+(x\ln x)^3}\;dx$

c) $\displaystyle\int (1+\cos x)\sqrt{1-(x+\sin x)^2}\;dx$

d) $\displaystyle\int\left(1-\frac{2}{x^3}\right)\sqrt{1-\left(x+\frac{1}{x^2}\right)^2}\;dx$

e) $\displaystyle\int x(1+x)\,e^{2x}\sin\sqrt{x e^x}\;dx$

f) $\displaystyle\int x^x\sqrt{x\ln x}\,(1+\ln x)\,dx$

g) $\displaystyle\int \frac{x\sin x\ln(x\sin x)}{1-\sqrt{1-\sqrt{x\sin x}}}(x\cos x+\sin x)\,dx$

h) $\displaystyle\int (1+\ln x)\sqrt{1-(x\ln x)^2}\;dx$

9.11 The Brightest Phase of Venus

Background: Trigonometry; optimization; area.

The brightness of Venus, as seen from Earth, is proportional to its visible area and inversely proportional to the square of the distance from Venus to Earth. In the figure below, note that as the angle ϕ increases from 0 to π, the visible area increases, thus increasing Venus's brightness. But the distance d from Earth to Venus also increases, which tends to decrease the brightness of Venus. Venus will appear brightest from Earth for some value of ϕ between 0 and π.

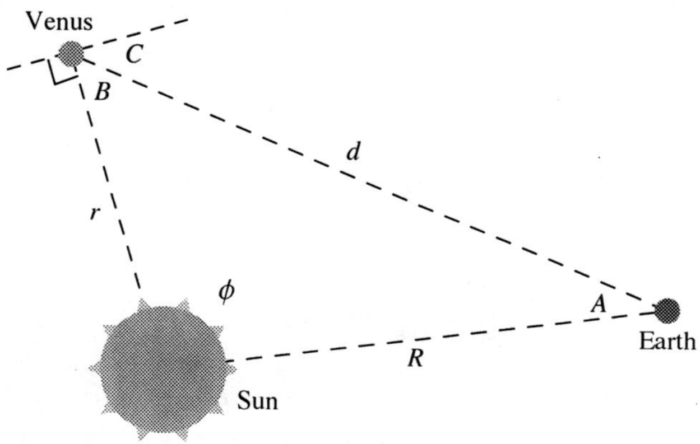

Our goal is to find the brightest "phase" of Venus by determining the angle ϕ for which Venus appears brightest from Earth. In the following, angles A, B, C, and ϕ and distances r, R, and d are as labeled in the figure above.

1. Use basic geometry to conclude that $B = \pi - (\phi + A)$ and $C = \phi + A - \pi/2$.
2. Let a be Venus's radius. Show that the visible portion of Venus lies between the curves $x = -\sqrt{a^2 - y^2}$ and $x = \sqrt{a^2 - y^2}\,\sin C$. Then show that the visible area of Venus is
$$\pi a^2(1 + \sin C)\big/2.$$
3. Let β represent the brightness of Venus as viewed from Earth. As described in the overview above, β is proportional to the quantity
$$\frac{\text{visible area}}{d^2}.$$
Use this, together with the results of #1 and #2, to find a formula for β in terms of the angles ϕ and A and the distance d. Then use the Law of Cosines and the Law of Sines from trigonometry to show that
$$d^2 = r^2 + R^2 - 2rR\cos\phi \text{ and } \sin A = \frac{r}{d}\sin\phi,$$
where r is the distance from the Sun to Venus, and R the distance from the Sun to Earth.
4. Use the equations in #3 with the values $R = 93$, $r = 67$, and $a = 0.004$ (each in millions of miles) to express β in terms of ϕ. Then find the angle ϕ between 0 and π that maximizes β.

9.12 The Skimpy Donut

Background: Volume of a solid of revolution; area of a surface of revolution; optimization.

The GetFat Donut Company makes donuts with a layer of chocolate icing. The company wants to cut costs by using as little chocolate icing as possible without changing the thickness of the icing or the weight (i.e., volume) of the donut. The problem, then, is to determine the dimensions of the donut that will minimize its surface area.

Assume that the donut has the idealized shape of a torus—obtained by revolving a circle about its central axis—as shown below.

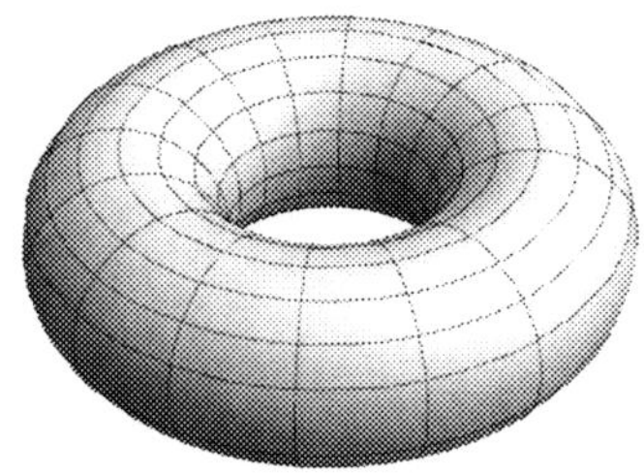

Such a torus can be obtained by revolving the circle $(x - a)^2 + y^2 = b^2$ about the y-axis, where b is the radius of the revolved circle, and a is the distance from the center axis to the center of the revolved circle. Previously, the donuts have been in the shape of such a torus with $a = 1$ inch and $b = 1/2$ inch.

1. Use the cylindrical shells technique to find the volume of the torus in terms of a and b. (Note that because of symmetry, it suffices to double the volume of the top half.) Now compute the volume V_0 that the donuts have had with dimensions $a = 1$ inch and $b = 1/2$ inch.
2. Find the surface area of the torus in terms of a and b. (Note that because of symmetry, it suffices to double the surface area of the top half.) As a check, the surface area should be $2\pi^2$ when $a = 1$ and $b = 1/2$.
3. With the volume fixed at its previous value V_0, determine the range of allowable values for each of a and b. (One important condition comes from the geometry of the torus.)
4. With the volume fixed at its previous value V_0, find the dimensions a and b that will minimize the donut's surface area.
5. Is there a *maximum* surface area for a fixed volume?

9.13 Designing a Light Bulb

Background: Slope, continuity, and differentiability; surfaces of revolution, optimization; the definite integral; systems of linear equations; using `Piecewise`.

The glass portion of a light bulb is to be in the form of a surface of revolution obtained by revolving the curve shown in the figure about the x-axis. The radius of the bulb's base is 1/2 inch, and the radius of its spherical portion is 1.25 inches.

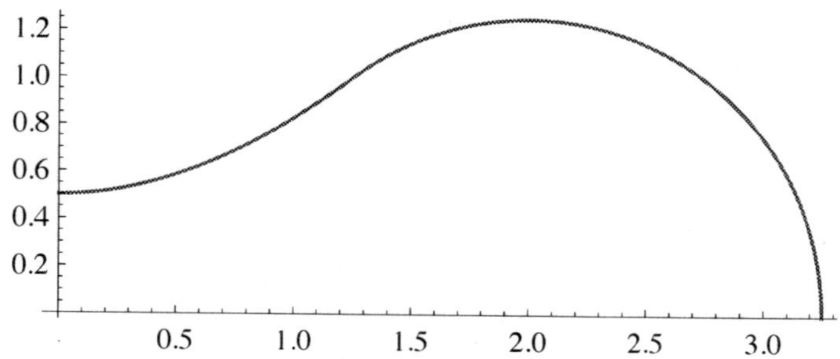

The curve is the graph of the piecewise-defined function

$$f(x) = \begin{cases} a\,x^3 + b\,x^2 + c\,x + d, & \text{when } 0 \le x \le 1.25 \\ \sqrt{1.25^2 - (x-2)^2}\,, & \text{when } 1.25 \le x \le 3.25 \end{cases}$$

1. Find the coefficients a, b, c, and d. Then plot the graph of f over $0 \le x \le 3.25$. (*Hint*: At each of $x = 0$ and $x = 1.25$, there are two conditions given by i) the point on the curve and ii) the slope of the curve.)

2. Use a `Piecewise` to define $f(x)$ as `f[x]`. Then enter the following to create a 3D plot.

```
ParametricPlot3D[{x, f[x] Sin[θ], f[x] Cos[θ]}, {x, 0, 3.25},
 {θ, 0, 2 π}, Mesh → None,
 PlotStyle → {Glow[Black], Opacity[.75]}, Exclusions → None,
 Axes → False, Lighting → "Neutral", Background → Darker[Blue]]
```

3. This problem concerns the placement of the filament in the bulb. The filament will be located at a point P on the central axis of the bulb, r inches from the base, such that the distance from the point P to the entire inner surface of the bulb, as measured by

$$\phi(r) = \int_0^{3.25} \sqrt{(r-x)^2 + f(x)^2}\, dx,$$

is minmized. Find the value of r that minimizes $\phi(r)$. Examine a standard light bulb and notice whether this result is close to what the bulb manufacturer actually uses. Also compare the bulb's profile to the curve in the figure above.

4. This problem is intended to shed some light on the approach to placing the bulb's filament used in #3. Show that the distance from a point $P = (r, 0)$ to the semicircle $y = \left(R^2 - x^2\right)^{1/2}$, as measured by

$$\phi(r) = \int_{-R}^{R} \sqrt{(r-x)^2 + R^2 - x^2}\, dx,$$

is minimized by $r = 0$; i.e., when P is the center of the circle.

9.14 Approximate Antidifferentiation

Background: Antidifferentiation; the Fundamental Theorem of Calculus; approximate integration; the trapezoidal rule.

The Fundamental Theorem of Calculus tells us that every continuous function f on an interval $[a, b]$ has an antiderivative $F(x)$, namely

$$F(x) = \int_a^x f(t)\,dt.$$

In many cases, such as when $f(x) = \cos x, x^3, e^x, x \sin x^2$, and so on, $F(x)$ can be express in terms of other elementary functions. However, for many (in fact, most) functions, there is no simpler way of expressing an antiderivative. Our goal here is to use the trapezoidal rule to develop an efficient method for approximating $F(x)$.

Preliminaries: Note that if we have a value for $F(x)$, then we can express $F(x+h)$ as

$$\begin{aligned} F(x+h) &= \int_a^x f(t)\,dt + \int_x^{x+h} f(t)\,dt \\ &= F(x) + \int_x^{x+h} f(t)\,dt. \end{aligned}$$

A two-point trapezoidal rule approximation to this last integral gives us

$$F(x + \Delta x) \approx F(x) + \frac{h}{2}\,(f(x) + f(x+h)).$$

Now, using this approximation, we can approximate values of F quite efficiently at a sequence of points $x_1, x_2, x_3, \ldots$, where $x_n = a + nh$, by using the recurrence formula

$$y_{n+1} = y_n + \frac{h}{2}\,(f(x_n) + f(x_{n+1})) \quad \text{for } n = 0, 1, 2, \ldots,$$

where y_i represents the resulting approximation to $F(x_i)$.

To implement the method in *Mathematica*, first define

```
step[{x_, y_}] := {x + h, y + .5 h (f[x] + f[x + h])}
```

Then after defining the function `f` and the "stepsize" `h`, `NestList` can be used as follows to compute points along the graph of the antiderivative. For example:

```
f[x_] := Cos[π Sin[π x]];
h := 2. / 100;
a := 0;
points = NestList[step, {a, 0}, 100];
```

The resulting points can be plotted with `ListPlot` as follows.

```
ListPlot[points, PlotJoined → True]
```

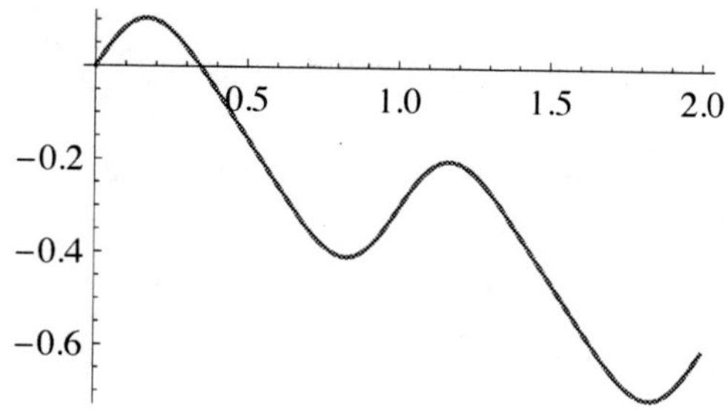

1. Reproduce the example above.
2. For each of the following, first plot the given antiderivative on the specified interval. Then plot the integrand on the same interval and `Show` the two graphs on the same plot. For each pair of graphs, discuss the two expected relationships:

   i) $\int_a^x f(t)\,dt$ is the "net" area between the graph of f and the x-axis between a and x;

   ii) $f(x)$ is the derivative of $\int_a^x f(t)\,dt$.

   a) $\int_0^x \sin t^2\,dt$, for $0 \le x \le 5$

   b) $\int_1^x \sin t^2\,dt$, for $1 \le x \le 5$

   c) $\int_0^x e^{-t^2}\,dt$, for $0 \le x \le 3$

   d) $\int_0^x e^{\sin t}\,dt$, for $0 \le x \le 4\pi$

   e) $\int_0^x \sin(t \cos t)\,dt$, for $0 \le x \le 2\pi$

   f) $\int_0^x \sin(t + \cos t)\,dt$, for $0 \le x \le 4\pi$

3. Modify the procedure described above so that it uses a three-point Simpson's Rule approximation rather than a two-point Trapezoidal approximation. Rework 2a and 2e above to test your modification.

9.15 Percentiles of the Normal Distribution

Background: Approximate integration; improper integrals; Newton's method.

The function

$$g(x) = \frac{1}{\sqrt{2\pi}}\, e^{-x^2/2}$$

is very important in Probability. Its graph is the standard normal (or bell) curve. If x represents an observable, random quantity (a random variable) that is "normally distributed" with mean value zero and standard deviation 1, then the area under the graph of g between $x = a$ and $x = b$ is the probability that a random observation of x will fall between a and b. If we define the notation $P(a \le x \le b)$ to represent this probability, we can then write

$$P(a \le x \le b) = \int_a^b g(x)\,dx.$$

Also,

$$P(x \le b) = \int_{-\infty}^{b} g(x)\,dx \text{ and } P(x \ge a) = \int_{a}^{\infty} g(x)\,dx$$

Our goal here is to construct a short table of standard *percentiles* of the normal distribution— specifically, a table of values of b corresponding to

$$P(x \le b) = .01,\ .05,\ .10,\ .25,\ .333,\ .50,\ .667,\ .75,\ .90,\ .95,\ .99.$$

1. Plot the graph of g over $-3.5 \le x \le 3.5$. Then compute numerical values for each of:

   a) $\int_{-1}^{1} g(x)\,dx$ b) $\int_{-2}^{2} g(x)\,dx$ c) $\int_{-3}^{3} g(x)\,dx$

2. Confirm by direct calculation that $\int_{-\infty}^{\infty} g(x)\,dx = 1$. Since g is an even function, it must be the case that $\int_{-\infty}^{0} g(x)\,dx = \frac{1}{2}$. Confirm this by direct calculation as well.

3. Let's try first to find the 75th percentile, that is, the value of b such that

$$\int_{-\infty}^{b} g(x)\,dx = .75$$

Since $\int_{-\infty}^{0} g(x)\,dx = \frac{1}{2}$, we can avoid the improper integral by finding the value of b such that

$$\int_{0}^{b} g(x)\,dx = .25$$

a) By trial and error, find a three decimal-place approximation to b.

b) Think of the current problem as that of finding a zero of the function

$$f(b) = \int_{0}^{b} g(x)\,dx - .25\,.$$

Using your approximation from (a) as an initial guess, perform one step of Newton's method to refine the approximation. Report the result rounded to five decimal places. (Hint: $f'(b) = g(b)$ by the Fundamental Theorem of Calculus.)

4. Repeat the process in #3 to find values of b corresponding to

$$\int_{-\infty}^{b} g(x)\,dx = .667,\ .90,\ .95,\ \text{and } .99\,.$$

5. Using symmetry and the percentiles already found, find the values of b for which

$$\int_{-\infty}^{b} g(x)\,dx = .01,\ .05,\ .10,\ .25,\ \text{and } .333\,.$$

9.16 Equilibria and Centers of Gravity

Background: Using the definite integral to compute arc length and centers of gravity; tangent and normal lines to a curve.

Suppose that you place a hard-boiled egg on a flat surface. It might wobble about a bit, but it will soon come to rest in a stable position. Also, if you tried hard enough, you could probably come very close to balancing the egg on either of its two ends. All of these balanced or rest positions are equilibrium positions. Two natural questions to ask are: (i) what makes a position an equilibrium position, and (ii) why is one equilibrium position stable and another not? In this project we will investigate these issues in a simple case.

1. A metal plate, with uniform thickness and mass density, occupies the region bounded by the parabola $y = x^2/4$ and the line $y = 4$ when balanced on its vertex. By symmetry, the center of gravity of the plate is at a point $(0, \overline{y})$ on the y-axis. Find $\overline{y}$.

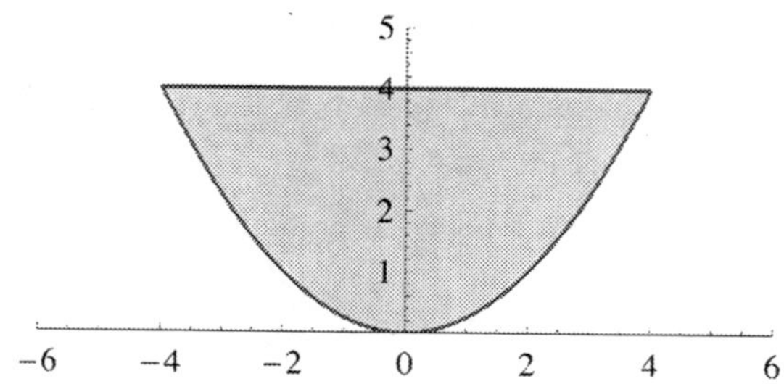

2. Think of the x-axis as the surface of a table, so that y is the vertical distance above the table. When the plate is balanced on its vertex at the origin, it is in an unstable position, because if the plate is disturbed ever so slightly, it will roll to one side and come to rest at a new equilibrium position. Suppose that the plate tips over to the right. Describe the new equilibrium position; in particular, find the new point where the plate touches the table and the new coordinates of its center of gravity. Intuitively, why is this position stable, while the original position was not?

3. Suppose that the plate occupies the region bounded by the parabola $y = x^2/4$ and the line $y = 9/4$ when balanced on its vertex. Show that there is no other equilibrium position for this plate. Conclude that the vextex-balanced position must be stable. (Why?)

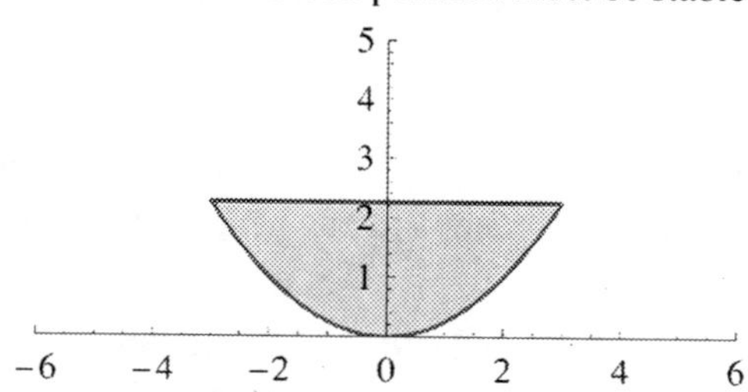

4. Suppose that the plate occupies the region bounded by the parabola $y = x^2/4$ and the line $y = b$ $(b > 0)$ when balanced on its vertex. Find the greatest value of b for which being balanced on its vertex is the plate's only equilibrium position.

5. Give simple geometric arguments for: a) why a semicircular plate has only one equilibrium position, and b) why every position of a circular plate is an equilibrium position.

9.17 Draining Tanks

Background: Volume; separable differential equations.

Consider a tank whose horizontal cross-sectional area is described by $A(y)$, where y is the vertical distance to the bottom of the tank. For example, a tank in the shape of a right circular cylinder would have $A(y) = \pi r^2 = \text{constant}$, and a tank in the shape of a cone (with vertex pointing down) would have $A(y) = \pi (\tan(\phi/2)\, y)^2$, where ϕ is the interior angle of the cone's central vertical cross-section at the vertex.

Toricelli's Law states that the rate at which a fluid drains through a hole in the bottom of the tank is proportional to the square root of the depth of the fluid; that is,

$$V'(t) = -k\sqrt{y},$$

where k is a positive constant that depends upon both the viscosity of the fluid and the size of the hole in the bottom of the tank. Since $V(y) = \int_0^y A(u)\,du$ and consequently $V'(t) = A(y)\,y'(t)$, Toricelli's Law gives us the differential equation

$$A(y)\,y'(t) = -k\sqrt{y},$$

the separated form of which is

$$A(y)\,dy \Big/ \sqrt{y} = -k\,dt.$$

1. A tank has the shape of an upright right circular cylinder with radius $R = 1$ foot and height $H = 3$ feet. The tank is initially full of water and begins to drain through a hole in the bottom at time $t = 0$. Assuming that $k = 0.5$, find the depth of water $y(t)$ in the tank for $t \geq 0$. How much time does it take for the tank to empty?
2. A tank has the shape of a cone (with vertex pointing down), where the interior angle of the central vertical cross-section at the vertex is $\phi = \pi/3$. The tank is initially filled with oil to a depth of 3 feet and begins to drain through a hole in the bottom at time $t = 0$. Given that $k = 0.1$, find the depth $y(t)$ of oil in the tank for $t \geq 0$. How much time does it take for the tank to empty?
3. A tank with height 4 feet has the shape of the solid obtained by revolving the graph of $y = x^2$ about the y-axis. It is initially full of water and begins to drain through a hole at its vertex at time $t = 0$.
   a) Find, in terms of the constant k, the depth of water $y(t)$ in the tank for $t \geq 0$. How much time (in terms of k also) does it take for the tank to empty?
   b) You observe that the tank empties in 4.8 minutes. What then is the value of k?

If the initial depth of the fluid in the tank is y_0, then the time T that it takes for the tank to empty satisfies

$$\int_{y_0}^{0} A(y) \Big/ \sqrt{y}\ dy = -k \int_0^T dt$$

from which follows easily the formula

$$T = k^{-1} \int_0^{y_0} A(y) \Big/ \sqrt{y}\ dy.$$

4. Consider a tank in the shape of a right circular cylinder with height H and radius R. Find the time required for the tank, initially full, to completely drain if:

   a) the tank is upright (i.e., flat end down); b) the tank lies on its side.

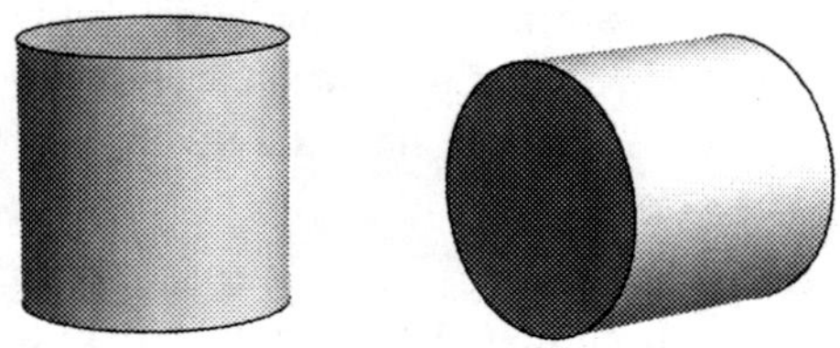

5. For the tank in #4 suppose that $H = 2R$, so that the initial depth is the same for each of the two positions. In which of the two positions will the tank drain faster?

9.18 Spruce Budworms

Background: First-order differential equations.

Spruce budworms are a serious problem in Canadian forests. Budworm "outbreaks" can occur in which balsam fir trees are quickly defoliated by hordes of ravenous budworms. A model of a budworm population leads to the differential equation

$$y'(t) = y\,f(y) \quad \text{with} \quad f(y) = \frac{k}{10}(10 - y) - \frac{y}{1 + y^2},$$

where y represents some fixed fraction (e.g, one one-thousandth) of the number of budworms congregated in a certain area, and k is a positive parameter associated with birth and death rates. An "outbreak-threshold" value of y is a positive number y^* such that:

$y(t)$ is decreasing for $t > 0$ if $y(0) \approx y^*$ and $y(0) < y^*$;

$y(t)$ is increasing for $t > 0$ if $y(0) \approx y^*$ and $y(0) > y^*$.

The roots of f, which depend upon the value of k and correspond to positive equilibrium solutions, determine the outbreak threshold, should one exist.

1. Graph $y\,f(y)$ versus y for a few values of k between 0.1 and 1. What are the possibilities for the number of positive roots?
2. For each of $k = .3$, $.45$, and $.6$, plot the direction field of the differental equation for $0 \le t \le 20$ and $0 \le y \le 10$. Also, use Euler's method to plot approximate solution curves with initial values $y(0) = 1$, 3, and 10. What happens to each of the solutions as $t \to \infty$?

3. For which of $k = .3$, $.45$, and $.6$ is an outbreak possible? What is the outbreak-threshold value of y in each case?
4. To find the values of k for which f has a "double root," solve simultaneously the equations $f(y) = 0$ and $f'(y) = 0$ for k and y. Then, using the resulting values of k, repeat #2.
5. State precisely the values of k for which a budworm outbreak is possible.

9.19 Parachuting

Background: First-order differential equations.

A sky-diver, weighing 70 kg, jumps from an airplane at an altitude of 700 meters and falls for T_1 seconds before pulling the rip cord of his parachute. A landing is said to be "gentle" if the velocity on impact is no more than the impact velocity of an object dropped from a height of 6 meters.

The distance that the sky-diver falls during t seconds can be found from Newton's Second Law, $F = ma$. During the free-fall portion of the jump, we will assume that there is essentially no air resistance—so $F = -mg$, where $g \approx 9.8$ meters/sec^2 and $m = 70$ kg. After the parachute opens, a significant "drag" term due to the air resistance of the parachute affects the force F, causing the force to become

$$F = -mg - kv,$$

where v is the velocity, and k is a positive drag coefficient. Assume that $k = 110$ kg/sec.

1. Find the range of impact velocities that correspond to a gentle landing.
2. Find the height and velocity after T_1 seconds of free-fall.
3. Find the impact velocity as a function of T_1.
4. Find the range of times T_1 at which the rip cord can be pulled for a gentle landing.

9.20 The Flight of a Baseball I

Background: Parametric equations; velocity and acceleration; antidifferentiation; optimization.

When a baseball player hits a fly-ball, many very complex physical phenomena affect its path—such as forces resulting from the spin of the ball, wind currents, and air resistance. However, we can learn a lot from considering a couple of idealized models of the path of the ball. The first of these ignores all forces other than gravitational force.

Suppose that, at the instant when it is hit, the ball has an initial speed v_0, the tangent line to its path forms an angle ϕ with the ground, and its height above the ground is y_0. Let $x = x(t)$ and $y = y(t)$ be the parametric equations of the path, and assume that $x(0) = 0$ and $y(0) = y_0$. The only force on the ball is the downward vertical force of gravity. Thus we can write accelerations in each direction as

$$x''(t) = 0 \quad \text{and} \quad y''(t) = -g.$$

The initial velocities in the two directions are $x'(0) = v_0 \cos\phi$ and $y'(0) = v_0 \sin\phi$.

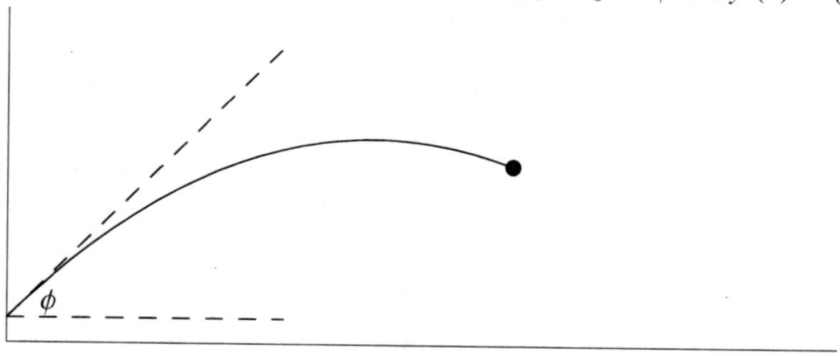

1. Show that integration of the accelerations results in

$$x'(t) = v_0 \cos\phi \quad \text{and} \quad y'(t) = -g\,t + v_0 \sin\phi,$$

and that integrating each of these once more produces

$$x(t) = (v_0 \cos\phi)\,t \quad \text{and} \quad y(t) = -\frac{g}{2}\,t^2 + (v_0 \sin\phi)\,t + y_0.$$

2. Let's assume that distance units are feet and time units are seconds, so that g may be approximated as 32 feet/sec^2. Then the parametric equations for the path of the ball may be entered as follows.

```
x[t_] := v0 Cos[ϕ] t;  y[t_] := -16 t^2 + v0 Sin[ϕ] t + y0
```

Suppose that $v_0 = 100$ feet/sec, $\phi = \pi/6$, and $y_0 = 3$ feet. Begin by plotting the path with `ParametricPlot` for $0 \le t \le 2$ with `PlotRange→{{0,300},{0,60}}` and then adjust the end-value of t in order to estimate:

a) how many seconds the ball remains aloft;

b) the distance from home plate to where the ball hits the ground.

3. After clearing the variables `v0`, ϕ, and `y0`, find—in terms of v_0, ϕ, and y_0—the time t_{fin} at which the ball hits the ground (assuming, of course, that $v_0 > 0$, $0 \le \phi \le \pi/2$, and $y_0 \ge 0$). Then find—also in terms of v_0, ϕ, and y_0—the distance x_{fin} from home plate to where the ball hits the ground.

4. Using $y_0 = 3$, find the angle ϕ that maximizes x_{fin} for each of $v_0 = 20,\ 50,\ 100,\ 150$, and 200. Convert each result to degrees.

5. Suppose that the home run fence is 20 feet high and 350 feet from home plate. Assuming that $y_0 = 3$ feet, find the minimum initial velocity with which the ball must be hit in order to clear the fence. Approximate the answer graphically (i.e., experimentally), and then find it by solving a pair of equations. (*Suggestion*: Use `PlotRange→ {{0,350},{0,100}}`, and `AxesOrigin→{350,0}`.)

6. Suppose that a certain player is capable of hitting the ball with an initial velocity no greater than 125 feet per second. Assuming that he does hit the ball with that initial velocity, what is the smallest angle ϕ that will carry the ball over a 10 foot fence that is 400 feet from home plate? Again, use $y_0 = 3$ feet. Approximate the answer graphically (i.e., experimentally), and then find it by solving an equation or pair of equations.

9.21 The Flight of a Baseball II

Background: Parametric equations; velocity and acceleration; antidifferentiation; optimization; the previous project, **The Flight of a Baseball I**.

By including air resistance in the model, we can obtain a much more realistic picture of the path of a baseball. A simple way of including air resistance in the model is to assume that it is a force that acts in the direction opposite that of the path and is proportional to the speed of the ball. The equations for the accelerations (in feet/sec^2) then become

$$x''(t) = -k\,x'(t) \text{ and } y''(t) = -32 - k\,y'(t),$$

where k is a "drag coefficient" divided by the mass of the ball. *Mathematica*'s `DSolve` can be used to solve for x and y as follows.

```
Clear[k, x, y, v0, ϕ];
sol = DSolve[{x''[t] == -k x'[t], y''[t] == -32 - k y'[t], x[0] == 0,
    x'[0] == v0 Cos[ϕ], y[0] == y0, y'[0] == v0 Sin[ϕ]}, {x, y}, t];

{x[t_], y[t_]} = {x[t], y[t]} /. sol[[1]] // Simplify
```

$$\left\{\frac{\text{v0 Cos}[\phi] - e^{-k t}\,\text{v0 Cos}[\phi]}{k},\ \frac{32 - 32\,e^{-k t} - 32\,k\,t + k^2\,\text{y0} + \left(1 - e^{-k t}\right) k\,\text{v0 Sin}[\phi]}{k^2}\right\}$$

1. Enter the commands above. Then enter the parametric equations from the previous project, **The Flight of a Baseball I**, as follows.

   ```
   xVac[t_] := v0 Cos[ϕ] t; yVac[t_] := -16 t^2 + v0 Sin[ϕ] t + y0
   ```

   Then with $v_0 = 100\,\text{ft/sec}$, $\phi = \pi/6$, $y_0 = 3$ feet, and $k = 0.005\,\text{sec}^{-1}$, plot both paths (i.e., with and without air resistance). Are the two paths significantly different? Repeat with $k = 0.01, 0.05, 0.1,$ and 0.5.

2. Let $v_0 = 100\,\text{ft/sec}$, $\phi = \pi/6$, $y_0 = 3$ feet, and $k = 0.1\,\text{sec}^{-1}$. By graphical experimentation, find an estimate for the value of ϕ that maximizes the distance from home plate to where the ball hits the ground. Two-decimal-place accuracy is sufficient. Repeat with $k = 0.2$. How different are the two results?

3. Suppose that the home run fence is 20 feet high and 350 feet from home plate. Find the minimum initial velocity with which the ball must be hit in order to clear the fence. Use $k = 0.1$ and $y_0 = 3$ feet. Approximate the answer graphically (i.e., experimentally). Be careful here—the angle ϕ must be taken into account. *Suggestion*: First fix ϕ at the value found in #2 and estimate the necessary initial velocity. Then increase and decrease ϕ by a small amount, say 0.02 (about 1°), and decrease your estimate of v_0 if necessary. Repeat until you have v_0 to two-decimal-place accuracy. Compare the result to your answer to problem 5 in **The Flight of a Baseball I**.

4. Suppose that a certain player is capable of hitting the ball with an initial velocity no greater than 125 feet per second. Assuming that he does hit the ball with that initial velocity, what is the smallest angle ϕ that will carry the ball over a 10 foot fence that is 400 feet from home

plate? Again, use $y_0 = 3$ ft and $k = 0.1\ \text{sec}^{-1}$. Approximate the answer graphically (i.e., experimentally). Compare the result to your answer to problem 6 in **The Flight of a Baseball I.**

5. Air density, which depends upon altitude above sea level, affects the value of k significantly. In particular, the value of k in Denver may be about 10% smaller than the value of k in Miami. Suppose that $k = 0.09$ in Denver and $k = 0.10$ in Miami. If a ball is hit with $y_0 = 3$ ft, $v_0 = 135$ ft/sec, and $\phi = \pi/6$, how much farther from home plate would it land in Denver than in Miami? (Estimate to the nearest foot.)

9.22 Cannonball Wars

Background: Parametric equations; velocity and acceleration.

One cannon is perched atop a cliff, 50 meters above the ground below. Another cannon is on the ground, 100 meters from the base of the cliff. (See the figure below.) The inclination of each cannon differs from the line of sight between them by the same angle α (as shown in the figure). Each cannon fires a cannonball at exactly the same instant with exactly the same initial velocity v_0 m/sec.

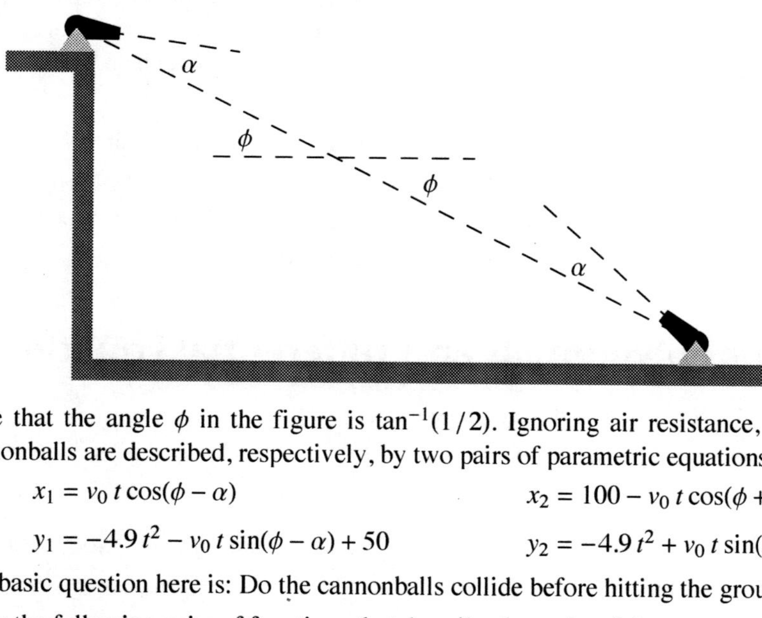

Note that the angle ϕ in the figure is $\tan^{-1}(1/2)$. Ignoring air resistance, the paths of the cannonballs are described, respectively, by two pairs of parametric equations:

$$x_1 = v_0\, t \cos(\phi - \alpha) \qquad x_2 = 100 - v_0\, t \cos(\phi + \alpha)$$

$$y_1 = -4.9\, t^2 - v_0\, t \sin(\phi - \alpha) + 50 \qquad y_2 = -4.9\, t^2 + v_0\, t \sin(\phi + \alpha)$$

The basic question here is: Do the cannonballs collide before hitting the ground?

1. Enter the following pairs of functions that describe the paths of the two cannonballs.

```
x1[t_] := v0 t Cos[ϕ - α];
y1[t_] := -49 t^2/10 - v0 t Sin[ϕ - α] + 50
x2[t_] := 100 - v0 t Cos[ϕ + α];
y2[t_] := -49 t^2/10 + v0 t Sin[ϕ + α]
```

Also enter

```
ϕ := ArcTan[1 / 2]
```

The following code produces a `Manipulate` panel for the cannonball trajectories in the case where $v_0 = 20$ m/sec and $\alpha = 15^\circ$. Enter it and experiment with the result. What do you observe? Do the cannonballs collide before hitting the ground?

```
v0 = 20; α = 15 Degree;
Manipulate[Show[ParametricPlot[
   {{x1[t], y1[t]}, {x2[t], y2[t]}}, {t, -.01, k}],
  Graphics[{PointSize[.025], Point[{x1[k], y1[k]}],
    Point[{x2[k], y2[k]}]}],
  PlotRange → {{0, 125}, {0, 92}} ] , {k, 0, 10}]
```

2. Keeping $v_0 = 20$, replot the paths with $\alpha = 30^\circ, 40^\circ$, and 50°. Do the cannonballs collide each time? What can be said about when they don't?
3. Change the value of v_0 to 40, and plot the paths for $\alpha = -15^\circ, 15^\circ, 45^\circ$, and 70°. Do the cannonballs collide each time? What can be said about when they don't?

In problems 4 and 5 below, assume that $\phi - \pi/2 < \alpha < \phi + \pi/2$ so that the cannon on the cliff fires its cannonball "forward." Also assume that $-\phi < \alpha < \pi - \phi$ so that the cannon on the ground fires its cannonball upward into the air.

4. Supposing temporarily that the paths of the cannonballs are never interrupted by any obstacle (such as the ground), show—by solving for it—that there is a positive time $t = t^*$ for which $x_1 = x_2$ and $y_1 = y_2$, i.e., a time when the two cannonballs are located at the same point.
5. For what values of v_0 and α does the collision of the cannonballs occur before one of them hits the ground? (*Suggestion*: Use `ContourPlot` to plot the curve $y_1(t^*) = 0$ for $-\pi/4 \le \alpha \le 3\pi/4$ and $0 \le v_0 \le 100$.) For what values of v_0 is a collision possible for some α? For what values of α is a collision possible for some v_0?

9.23 Taylor Polynomials and Differential Equations

Background: Taylor Series; basic differential equations.

Consider a first-order initial value problem

$$y'(t) = f(t, y), \;\; y(0) = y_0.$$

The initial condition obviously provides the value of the solution at $t = 0$. Notice also that the differential equation provides the value of $y'(t)$ at $t = 0$, namely

$$y'(0) = f(0, y_0).$$

Moreover, differentiation of each side of the differential equation with respect to t produces an expression for $y''(t)$,

$$y''(t) = \frac{\partial f}{\partial y}\, y'(t),$$

which can then be evaluated at $t = 0$, using known values of $y(0)$ and $y'(0)$. In principle, if this process is continued, we can determine the values at $t = 0$ of as many derivatives as we like and thereby obtain any Taylor polynomial of the solution about $t = 0$.

1. Consider the initial value problem

$$y'(t) = t^2 - y^2, \quad y(0) = 1.$$

   a) Create a list containing the first through sixth derivatives of y by entering

```
f[t_] := t^2 - y[t]^2;
deg := 6;
derivs = Join[{y[t]}, Table[D[f[t], {t, k - 1}], {k, 1, deg}]];
derivs // TableForm
```

   b) Evaluate the list of derivatives at $t = 0$ by entering

```
(derivs = derivs //. {t → 0, y[0] → 1}) // TableForm
```

   followed by

```
derivs = derivs //. (D[y[t], {t, k_}] /. t → 0) :> derivs[[k]]
```

   Now compute the coefficients of the desired Taylor polynomial by dividing by the appropriate factorials as follows.

```
coeffs = derivs / Table[k!, {k, 0, deg}]
```

   c) Define the nth degree Taylor polynomial $p_n(x)$ by entering

```
p[0][x_] := coeffs[[1]];
p[n_Integer][x_] := p[n - 1][x] + coeffs[[n + 1]] x^n /; n > 0
```

   Then create a list of the Taylor polynomials of degree one through six by entering

```
(polys = Table[p[n][x], {n, 1, deg}]) // TableForm
```

   d) Finally, plot the Taylor polynomials for $0 \le x \le 2$ by entering

```
Plot[Evaluate[polys], {x, 0, 2}, PlotRange → {-1, 1}]
```

2. Using the same process as in #1, plot the first- through sixth-degree Taylor polynomials of the solution of

$$y'(t) = \cos(t\,y), \quad y(0) = 0.$$

3. Using a *similar* process to that in #1, plot the first- through sixth-degree Taylor polynomials of the solution of the second-order problem

$$y''(t) = -t\,y, \quad y(0) = 1, \quad y'(0) = 0.$$

9.24 Build Your Own Cosine

Background: Taylor polynomials.

Suppose that you have a simple calculator whose only functions are the basic arithmetic functions: addition, subtraction, multiplication, and division. However the calculator does allow you to program your own functions, using the basic arithmetic functions as well as basic programming constructs such as *if*-statements and functions such as *floor* and absolute value. Your assignment here is to create a cosine function for your calculator. The function must compute $\cos x$ to within 5×10^7 for any x (thus giving six decimal-place accuracy).

Since the graph of $\cos x$ is periodic with period 2π, we first need to obtain an approximation on the interval $0 \le x \le 2\pi$. Also, because of symmetries in the graph of $\cos x$ on $0 \le x \le 2\pi$, we need only be concerned initially with obtaining an approximation on the interval $0 \le x \le \pi/2$. So it is sensible to consider Taylor polynomials about $x = \pi/4$, the midpoint of that interval.

1. Show that the eighth-degree Taylor polynomial for $\cos x$ about $x = \pi/4$ is the Taylor polynomial of least degree that will provide the accuracy we want on $0 \le x \le \pi/2$.

2. Compute a convenient form of the desired Taylor polynomial by entering

```
taylor8[x_] = Series[Cos[x], {x, π/4, 8}] // Normal // N // Factor
```

Note that the result could be easily programmed into your mythical calculator. Now plot this polynomial and $\cos x$ together over $0 \le x \le \pi/2$ and then over $-2\pi \le x \le 2\pi$.

3. Now that we have a good approximation to $\cos x$ for $0 \le x \le \pi/2$, let's extend that approximation to the interval $0 \le x \le \pi$ by means of the identity $\cos x = -\cos(\pi - x)$. Enter the following to obtain a good approximation to $\cos x$ for $0 \le x \le \pi$.

```
halfCos[x_] := If[x ≤ π/2, taylor8[x], -taylor8[π - x]]
```

Plot this function and $\cos x$ together over $0 \le x \le \pi$ and then over $-2\pi \le x \le 3\pi$.

4. Now that we have a good approximation to $\cos x$ for $0 \le x \le \pi$, let's extend that approximation to the interval $0 \le x \le 2\pi$ by means of the identity $\cos x = \cos(2\pi - x)$. Enter the following to obtain a good approximation to $\cos x$ for $0 \le x \le 2\pi$.

```
fullCos[x_] := If[x ≤ π, halfCos[x], halfCos[2 π - x]]
```

Plot this function and $\cos x$ together over $0 \le x \le 2\pi$ and then over $-2\pi \le x \le 4\pi$.

5. Now that we have a good approximation to $\cos x$ for $0 \le x \le 2\pi$, all we need to do is to build a function that will "shift" any x to its corresponding value between 0 and 2π. In the language of trigonometry, we simply need to find a coterminal angle between 0 and 2π for x radians. This is done by subtracting from x the greatest integer multiple of 2π that is less than or equal to x. This is accomplished by means of the `Floor` function. Enter

```
shift[x_] := x - 2 π Floor[x/(2 π)]
```

and plot the function over $0 \le x \le 8\pi$ with `AspectRatio→Automatic`. Finally, enter

```
myCosine[x_] := fullCos[shift[x]]
```

and plot the result over $-20 \le x \le 20$.

(*Note*: The function `shift[x]` is numerically equivalent to the built-in function `Mod[x,2π]`.)

Index

H

I

J

L

M

N

O

P

Q

R

S

T

U

V

W

Z